Yan Fan • Poh Seng Lee
Pawan Kumar Singh • Yong Jiun Lee

Thermal Transport in Oblique Finned Micro/Minichannels

Yan Fan
National University of Singapore
Singapore, Singapore

Poh Seng Lee
National University of Singapore
Singapore, Singapore

Pawan Kumar Singh
National University of Singapore
Singapore, Singapore

Yong Jiun Lee
National University of Singapore
Singapore, Singapore

ISSN 2191-530X ISSN 2191-5318 (electronic)
ISBN 978-3-319-09646-9 ISBN 978-3-319-09647-6 (eBook)
DOI 10.1007/978-3-319-09647-6
Springer Cham Heidelberg New York Dordrecht London

Library of Congress Control Number: 2014947519

Printed on acid-free paper

Springer is part of Springer Science+Business Media (www.springer.com)

SpringerBriefs in Applied Sciences and Technology

Thermal Engineering and Applied Science

Series editor
Francis A. Kulacki, Minneapolis, MN, USA

More information about this series at http://www.springer.com/series/10305

Preface

The continuous miniaturization of electronic devices, coupled with advanced packaging technology, led to the ever-increasing packaging densities and the associated heat flux that need to be dissipated. Thermal management has become a critical requirement for high-power heat sources, such as electronic chips, heat exchangers, high-performance battery packs and high-power motors. Thus, micro-/minichannel with oblique fins stands out as one of the most promising solutions. This design is created by cutting secondary channels at an oblique angle with the straight fins on the planar surface. It offers several advantages such as compactness, light weight and higher heat transfer surface area to fluid volume ratio, greater temperature uniformity across the heat transfer and diverse thermal management applications. Most interestingly, this structure was shown to enhance heat transfer performance significantly with negligible pressure drop penalty. This is different from the conventional passive heat transfer enhancement techniques where trade-off in terms of pressure drop penalty is always associated with the improved heat transfer.

The objectives of this brief on thermal transport in micro-/minichannel are to (1) propose a novel planar oblique fin microchannel and cylindrical oblique fin minichannel heat sink design using passive heat transfer enhancement techniques, (2) investigate the forced convection heat transfer and fluid flow characteristic in both planar and cylindrical oblique fin structures through 3D conjugate numerical simulation and systematic experimental studies, (3) examine the feasibility of employing the proposed oblique fin microchannel heat sink in cooling non-uniform heat fluxes and hotspot suppression, (4) examine and investigate the influence of edge effect on flow and temperature uniformity in cylindrical oblique fin heat sinks through systematic numerical and experimental studies, (5) perform a similarity analysis to obtain a dimensionless grouping parameter to evaluate the total heat transfer rate of the oblique fin heat sink, (6) investigate the flow mechanism and optimize the dimensions of cylindrical oblique fin heat sink for good overall heat transfer performance using similarity analysis and parametric numerical investigations and (7) obtain generalized correlations to predict the heat transfer performance and pressure drop characteristics of the oblique fin minichannel heat sink.

This brief may have significant impact on providing an innovative cooling solution and understanding the flow physics behind oblique fin structure. The novel oblique fin micro-/minichannel heat sink could enhance the heat transfer performance significantly and make the heat source temperature more uniform and not compromise with high pumping power. The ability to locally tailor heat transfer performance can be crucial for hotspot mitigation as the electronic industry is moving towards more-than-Moore technologies. The proposed technique could lead to a smaller and lighter cooling system and increase the longevity of the heat source and save substantially more energy. The findings from this work could aid in the design, optimization and implementation of oblique fin heat sinks.

Singapore, Singapore

Yan Fan
Poh Seng Lee
Pawan Kumar Singh
Yong Jiun Lee

Acknowledgments

The authors gratefully acknowledge the financial support received from the National University of Singapore and the MOE Academic Research Fund (AcRF) Tier 2 research project No.MOE2011-T2-2-126 (WBS No: R265-000-423-112) for their financial support for this work.

The editorial assistance of the staff at Springer is also much appreciated.

Contents

Abbreviations

Nomenclature

ΔP	Pressure drop, Pa
ΔT_m	Mean temperature difference between fluid and channel wall
A	Heat transfer surface area, mm^2
$C_{1,\ldots,10}$	Correlation factors
C_1, C_2	Undetermined constants
c_p	Specific heat capacity, KJ/Kg K
D	Heat source dimension, mm
D_1	Main channel width, mm
D_h	Hydraulic diameter, mm
F	Blending function
f	Friction factor
	Dimensionless stream function for Blasius solution
f_{app}	Apparent friction factor
h	Heat transfer coefficient, W/m^2 K
H	Channel height, mm
	Oblique fin width, mm
I	Current, A
K	Loss coefficient
k	Thermal conductivity, W/m K
	Turbulence kinetic energy
L	Total length of the heat sink, mm
l_f	Fin length, mm
l_{sc}	Secondary channel gap, mm
l_u	Characteristic length for one unit of oblique fin, mm
m, n	Exponents
n	The number of channel starting from draining edge
N	Number of channels
Nu	Nusselt number

P	Heat input, W
p	Perimeter, mm
	Pressure, Pa
P_k	Model production term
P_P	Pumping power, W
Pr	Prandtl number
Q	Volumetric flow rate, L/min
q	Heat gain by the fluid, W
q''	Heat flux, W/m^2
q_p	Heat flux per pump power and per temperature difference
q_s''	Surface heat flux, W/m^2
R	Thermal resistance, °C/W
R^2	Coefficient of determination
Re	Reynolds number
S	Absolute value of the shear strain rate
T	Temperature, °C or K
T_∞	Free stream or fluid temperature, K
T_s	Surface temperature, K
u	Average x-velocity, m/s
U	Voltage, V
u_1	Main flow velocity, m/s
u_2	Secondary flow velocity, m/s
u_∞	Free stream velocity, m/s
v	Average y-velocity, m/s
W	Main channel bottom width, mm
x, y, z	Cartesian coordinates
X, Y, Z	Nondimensional Cartesian coordinates

Greek Symbols

$\theta^{\sim}$	Dimensionless temperature difference
ω	Specific rate of turbulence dissipation
υ	Kinematic viscosity, m^2/s
σ	Diffusion parameters
Δ	Gradient
α	Aspect ratio
	Thermal diffusivity of fluid
	SST model constant
β	Edge angle
β_1, β_2	SST model constants
η	Fin efficiencySimilarity parameter
θ	Oblique angle, °
μ	Dynamic viscosity, Ns/m^2
ρ	Density, kg/m^3
Υ	Intermittency

Subscript

ave	Average
c	Contraction
ch	ChannelcuCopper
e	Expansion
eq	Equivalent
f	Fluid
fd	Fully developed
i, j	Tensor indices
in	Inlet
m	Mean bulk fluid
o	Outlet
t	Turbulent
tot	Total
w	Wall

Chapter 1
Introduction

Keywords Thermal management • Oblique finned micro-/minichannel • Secondary channel • Edge effect • Heat transfer • Pressure drop • Hotspot • Multiple correlations

1.1 Background

For more than 40 years, the number of transistors that can be placed inexpensively on an integrated circuit has doubled approximately every 18 months, following the famous Moore's law suggested by Gordon E. Moore [1]. This trend has led to radical evolvement in micro-fabrication technology since the invention of integrated circuit in the 1950s. Nowadays, transistors are made smaller than ever, while integrated circuits (chips) have much higher on-chip clock frequency than the previous generations. Unfortunately, the waste heat generated during the logic computation is also proportional to the clock frequency. Waste heat generated from electronics must be sufficiently removed to ensure that the operating temperature is kept within the optimum range. Once the temperature exceeds this range, the performance of the device would either degrade or its lifespan would shorten drastically. To make the problem worse, the emerging trend of product miniaturization eliminates the useful surface area for waste heat removal, leading to serious thermal management challenges. As early as 1975, Keyes [2] highlighted the severe thermal management threat posed by the continuous miniaturization of the electronics.

Thermal management, in the past, was considered as the secondary issue after product performance; however, this priority can no longer be assumed. In fact, Jenkins and Bennett [3] pointed out that thermal management has become the fundamental driver in the development of the architecture for advance designs. According to the analysis by Ramsey et al. in 2004 [4], 55 % of the failure that occurred in electronic systems was due to heat-related issues, prompting the importance of integrating thermal management considerations from the early stage of product design, to ensure the survival of overall system. Over the years, there

Y. Fan et al., *Thermal Transport in Oblique Finned Micro/Minichannels*,
SpringerBriefs in Applied Sciences and Technology,
DOI 10.1007/978-3-319-09647-6_1

has been extensive research performed to extend the heat removal capability of air-cooled heat sink to meet the fierce cooling requirements. Various areas such as dimensional optimization, radical fin design, thermal interface material, revolutionary fan design and hybrid heat sink have been explored [5–7], suggesting that air-cooled heat sinks were overstretched to keep up with the fierce requirements of electronic cooling.

The 2009 International Technology Roadmap for Semiconductors (ITRS) continues to predict a steep increment in the transistor density, on-chip clock frequency and power dissipation for future processors [8]. As a consequence, the use of highly effective and compact liquid cooling techniques is inevitable for the next generation of electronics [9]. Besides, liquid cooling is regarded as a potential candidate in cooling diode laser arrays (DLA) [10] and high-energy laser (HEL) weapons [11], which dissipate heat in the order of several kW/cm^2. Liquid (e.g. water), which has high thermal conductivity and specific heat capacity values, is far more effective than air at removing the waste heat. In addition, liquid cooling solution generates little noise, which is not achievable with active air cooling that involves the use of a fan. Furthermore, Garimella [12] and Agostini et al. [13] reviewed the various liquid cooling technologies available including their state-of-the-art systems. Among them, microchannel heat sink stands out as one of the most promising solutions. Tuckerman and Pease [14] in 1981 proposed microchannel liquid cooling technology and demonstrated that cooling electronic circuits with power density of more than 1,000 W/cm^2 should be feasible. In their study, they scaled liquid-cooled heat exchanger technology to microscopic dimensions and attached the compact heat sink onto a silicon chip. This idea realized two much appreciated elements in a thermal management solution: the significant increment in heat transfer area per unit volume and the heat transfer coefficient. The combined effect of increased heat transfer area and heat transfer coefficient lowers the thermal resistance by at least an order of magnitude compared to air cooling technology.

Although their work shed some light on a future thermal management solution, several drawbacks of the newly proposed microchannel concept were evident. Copeland et al. [15] revealed the application difficulties for the conventional microchannel, which were associated with high pressure drop and significant lateral temperature gradient. Due to the use of small size channels, pressure drop across the heat sink can be as high as two bars for the minimum total thermal resistance achieved. This pressure head can only be provided by a conventional pump, making it impractical for implementation in a small and portable system. In addition, as coolant flowed through the microchannel, the thermal boundary layer quickly developed and caused the heat transfer performance to deteriorate in the streamwise direction. This led to non-uniform heat transfer performance within the heat sink and elevated temperature at the downstream of the heat sink. As a result, a significant temperature gradient existed across the heat sink (electronic chip), inducing large thermal stress to the devices, which could possibly lead to reliability problems and failure. This limitation posed a significant risk to the technology advancement and subsequently sparked serious attention in microchannel research.

1.2 Objectives

The main aim of this work is to develop a novel and highly effective heat transfer augmentation technique for single-phase micro-/minichannel heat sink. This solution should enhance both local and overall heat transfer performance and eliminate the temperature maldistribution across the heat source surface while preventing significant pressure drop penalty. The specific objectives of this research are:

- Propose a novel oblique fin micro-/minichannel heat sink design for both planar and cylindrical heat source applications using passive heat transfer enhancement techniques.
- Investigate the forced convection heat transfer, fluid flow characteristic in the cylindrical oblique fin structure through 3D conjugate numerical simulation and systematic experimental studies to characterize the secondary flow effect in heat transfer and pressure drop in the micro-/minichannels.
- Examine the feasibility of employing the proposed enhanced microchannel heat sink in cooling non-uniform heat fluxes and hotspot suppression.
- Examine and investigate the influences of edge effect on flow and temperature uniformity in cylindrical oblique fin minichannel heat sinks through systematic numerical and experimental studies.
- Perform a similarity analysis to obtain a dimensionless grouping parameter to evaluate the total heat transfer rate of the heat sink.
- Investigate the flow mechanism and optimize the dimensions of cylindrical oblique fin heat sink for good overall heat transfer performance using similarity analysis and parametric numerical investigations.
- Obtain generalized correlations to predict the heat transfer performance and pressure drop characteristics of the cylindrical oblique fin minichannel heat sink when the parameter values are beyond those used in the parametric computations.

1.3 Scope

The results of this present study may have significant impact on both providing an innovative cooling solution for heat source application and understanding the flow physics behind oblique fin structure. The novel oblique fin micro-/minichannel heat sink could enhance the heat transfer performance significantly and make the heat source temperature more homogeneous and not compromise with high pumping power. The proposed technique could lead to a smaller and lighter cooling system and increase the longevity of the heat source and save substantially more energy. The ability to locally tailor heat transfer performance can be crucial for hotspot mitigation as the electronic industry is moving towards more-than-Moore technologies. The findings from this work could aid in the design, optimization and implementation of oblique fin micro-/minichannel heat sinks. Besides, the outcome of this research may be extended to air-cooled heat sink, which is another bottleneck in thermal management challenge.

The present research is only on laminar flow regime of single-phase liquid cooling. Air cooling and two-phase cooling are beyond the scope of this study. The heat transfer and fluid flow characteristics in these flow regimes are simulated by using the Navier–Stokes equation.

References

1. Moore GE (1965) Cramming more components onto integrated circuits. Electronics 38:114–117
2. Keyes RW (1975) Physical limits in digital electronics. Proc IEEE 63:740–766
3. Jenkins LC, Bennett A (2004) 21st century challenge: thermal management design requirements. 23rd digital avionics systems conference. DASC 04, vol 2, p 9. D. 1–7
4. Ramsey J, Jones KW, Mitra AK (2004) Thermal management solutions to advanced integrated and discrete bipolar junction (BJT) device structures. In: Proceeding of international conference of environmental system ICES 04, p 2004–2001–2572
5. Chan MA, Yap CR, Ng KC (2009) Modelling and testing of an advanced compact two-phase cooler for electronics cooling. Int J Heat Mass Transfer 52:3456–3463
6. Rodgers P, Eveloy V, Pecht MG (2005) Limits of air-cooling: status and challenges. Proceeding of 21st IEEE SEMI-THERM symposium
7. Sauciuc I, Chrysler G, Mahajan R, Szleper M (2003) Air-cooling extension: performance limits for processor cooling applications. Proceeding of 19th IEEE SEMI-THERM symposium
8. http://www.itrs.net/
9. Kandlikar SG (2005) High flux heat removal with microchannels: a roadmap of challenges and opportunities. Heat Transfer Eng 26:5–14
10. Huddle JJ, Chow LC, Lei S, Marcos A, Rini DP (2000) Thermal management of diode laser arrays. Proceeding of 16th IEEE SEMI-THERM symposium
11. Perram GP, Marciniak MA, Goda M (2004) High-energy laser weapons: technology overview. Proceedings of the SPIE: The Int Society for optical engineering, vol 5414
12. Garimella SV (2006) Advanced in mesoscale thermal management technologies for microelectronics. Microelectron J 37:1165–1185
13. Agostini B, Fabbri M, Park JE, Wojtan L, Thome JR, Michel B (2007) State of the art of high heatfluxcoolingtechnologies.HeatTransferEng28:258–281.doi:10.1080/01457630601117799
14. Tuckerman DB, Pease RFW (1981) High-performance heat sinking for VLSI. IEEE Electron Device Lett 2:126–129. doi:10.1109/EDL.1981.25367
15. Copeland D, Takahira H, Nakayama W (1995) Manifold microchannel heat sinks: theory and experiments. Therm Sci Eng 3:9–15

Chapter 2
Planar Oblique Fin Microchannel Structure

Keywords Planar oblique fin microchannel • Boundary layer • Secondary flow • Heat transfer • Pressure drop • Hotspot mitigation

2.1 Oblique Fin Concept and Motivations

In order to generate secondary flow in the conventional-sized passage, the louvred fin heat exchanger has the entire slit fins rotated 20°–45° relative to the airflow direction. On the other hand, Steinke and Kandlikar [1] suggested placing the smaller secondary channels at an angle between the main channels for microchannel application. The graphic representation of both concepts is illustrated in Figs. 2.1 and 2.2, respectively.

Both fin layouts share the same characteristics, where the secondary flow paths are rotated at an angle relative to the main flow paths, and these unique characteristics are adopted into the current microchannel heat sink design. The current heat sink design adopts sectional oblique fins to replace the conventional continuous fins in microchannel heat sink application [3]. Figure 2.3 illustrates the plan view of the proposed microchannel configuration along with the flow paths for the main flows and secondary flows.

In this design, oblique cuts are made along the fins to create smaller, branching secondary channels (named as oblique channel) with the intentions to (1) disrupt the thermal boundary layer development and (2) generate secondary flows. The current design would follow the recommendation by Suga and Aoki [4] that the ratio of fin pitch in the spanwise direction to the fin length has to be 1.5 times the tangent of oblique angle ($F_p/L_p = 1.5 \times \tan\,\theta$ as shown in Fig. 2.1) to provide a good balance between heat transfer and pressure drop. The oblique angle that denotes the angle between the main channel and oblique channel is set as ~27°, which is within the range of louvre angles (20°–45°) that are frequently evaluated in the literatures. Instead of emulating the design by Steinke and Kandlikar [1], where the direction of secondary channels is alternated in the streamwise direction, the direction oblique channel remains consistent throughout the heat sink, following the louvred fin design. The concept of turnaround louvre is, however, not incorporated.

Y. Fan et al., *Thermal Transport in Oblique Finned Micro/Minichannels*,
SpringerBriefs in Applied Sciences and Technology,
DOI 10.1007/978-3-319-09647-6_2

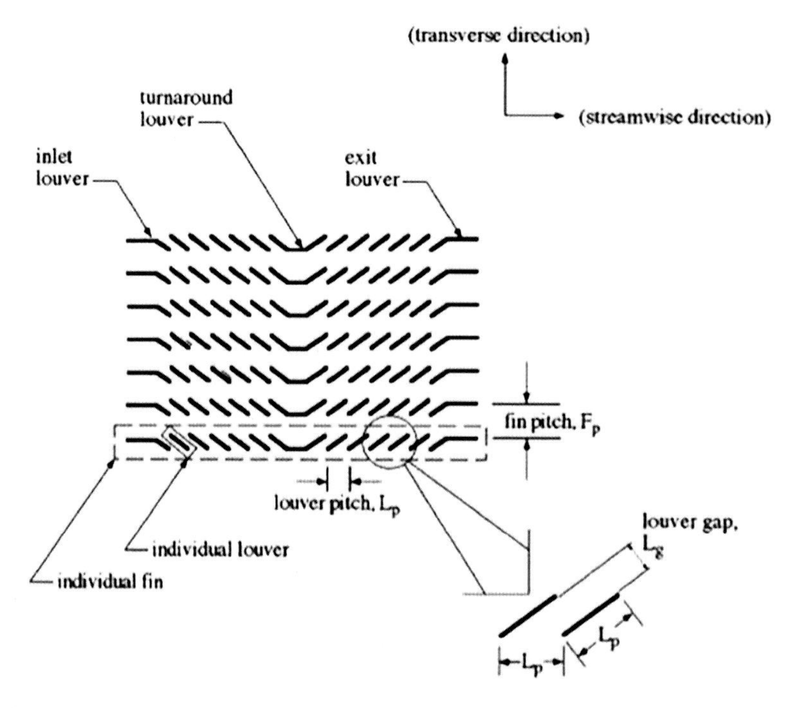

Fig. 2.1 Schematic of louvred fin arrays [2]

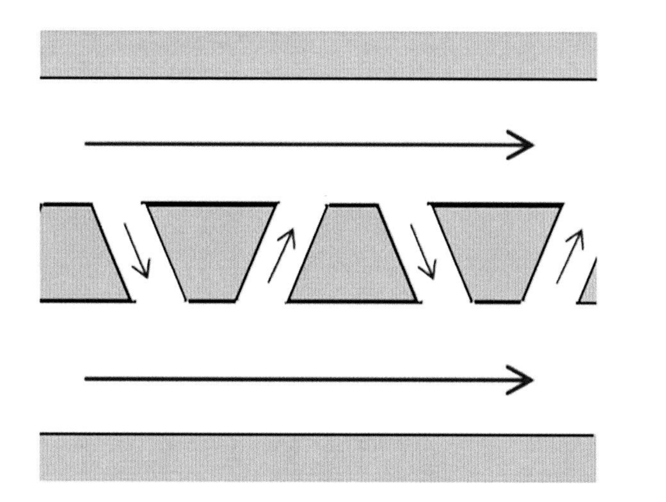

Fig. 2.2 Proposal by Steinke and Kandlikar to generate secondary flows [1]

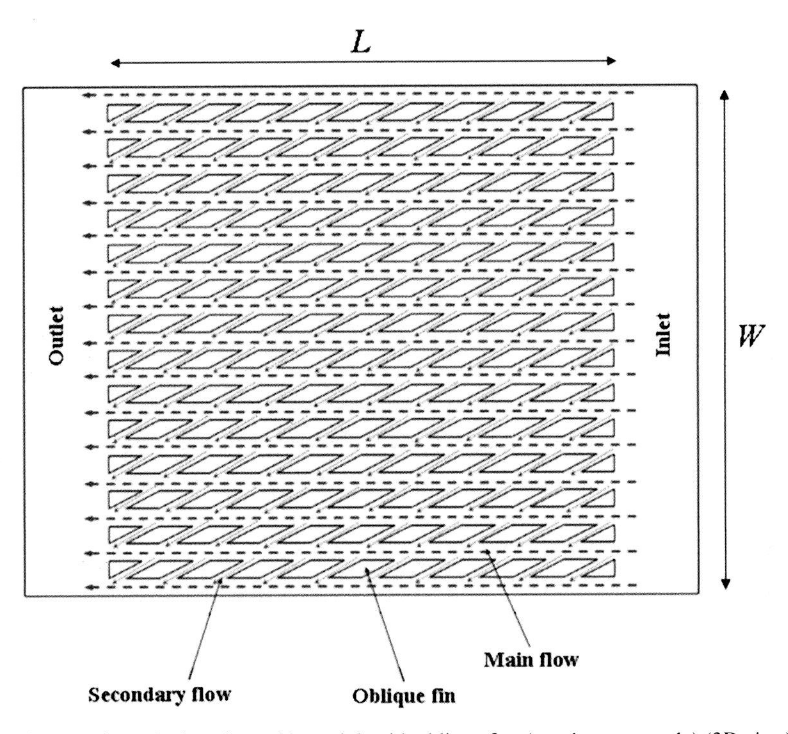

Fig. 2.3 Plan view of microchannel heat sink with oblique fins (not drawn to scale) (3D view) [3]

The width ratio of main channel to oblique channel is kept as 2:1, with the intention to reduce the amount of flow through secondary channel.

Although the current oblique fin design bears some resemblance to the louvred fin heat exchangers in the layout, the flow pattern in the enhanced microchannel with oblique fin array is designed to be different. Except at very low Reynolds number, the main flow within louvred fin arrays is nearly parallel to the louvres [5] and can be approximated as boundary layer flow, for which the Pohlhausen solution as flow over flat plate is applicable [6]. On the other hand, for the oblique fin array, the bulk of the flow remains in the main channel with only a small fraction of flow branching into the oblique openings (channels) and subsequently injected into the adjacent main channel, leading to generation of secondary flows which promote fluid mixing. For the sake of convenience and clarity, the microchannel with oblique fins in this chapter is addressed as enhanced microchannel.

2.2 Numerical Analysis: Simplified Model

The geometries for both the enhanced microchannel and conventional microchannel used in the simulation study are tabulated in Table 2.1. In order to facilitate a fair performance comparison, both heat sinks share the same channel aspect ratio, channel

Table 2.1 Dimensional details for microchannel heat sinks for simulation [3]

Characteristic	Conventional microchannel	Enhanced microchannel
Material	Copper	
Footprint, width × length (mm)	25 × 25	
Main channel width, w_{ch} (μm)	500	
Fin width, w_w (μm)	500	
Channel depth, H (μm)	1,500	
Aspect ratio, α	3	
Number of fin row, n	22	
Number of fin per row	–	12
Oblique channel width, w_{ob} (μm)	–	250
Fin pitch, p (μm)	–	2,000
Fin length, l (μm)	–	1,450
Oblique angel, θ (°)	–	27

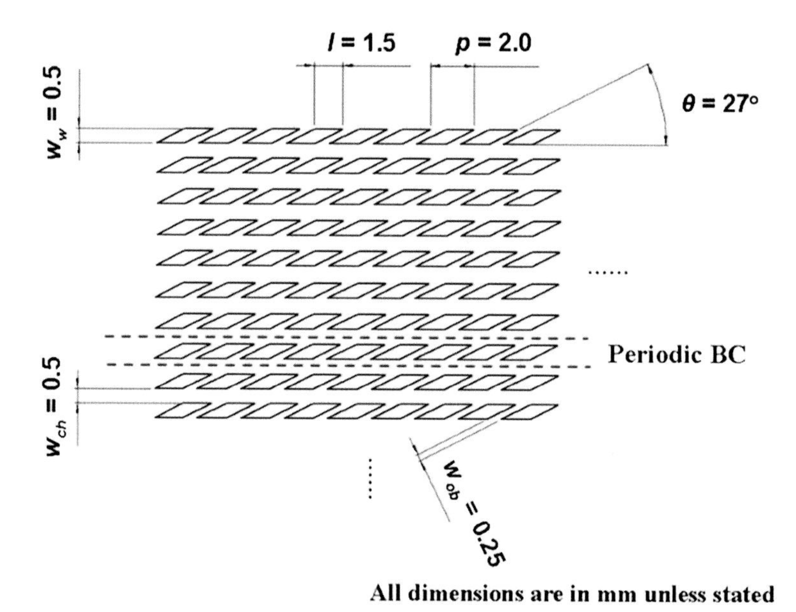

Fig. 2.4 Plan view of oblique fins with dimensions [3]

width, fin width and overall footprint. Apart from these common characteristics, the enhanced microchannel has openings that are obliquely cut at 27° from the main channel, at the pitch of 2 mm, with an oblique channel width of 250 μm. The dimensions for the enhanced microchannel are shown in Fig. 2.4.

2.2.1 CFD Simulation Approach

Both simulation domains for enhanced microchannel and conventional microchannel are generated using GAMBIT v2.3. As evident from Fig. 2.4, the oblique fin structure exhibits spanwise periodicity if the edge effect is neglected. Thus, periodic boundary condition is assumed to reduce the computation domain to a channel-fin pair consisting of full width oblique fins and associated oblique channels in the middle and two half-width main channels at the sides. The flow at a periodic boundary is treated as though the opposite periodic plane is a direct neighbour to the cells adjacent to the first periodic boundary. Thus, when calculating the flow through the periodic boundary adjacent to a fluid cell, the flow conditions at the fluid cell adjacent to the opposite periodic plane are used [7]. The enlarged view of the computational domain for enhanced microchannel is showed in Fig. 2.5.

The top surface of the fins and channels are first meshed with quad-pave scheme, with a 25 μm spacing spanwise and 25 μm spacing streamwise. The volume mesh for the main channel, oblique fins and oblique channels are then generated with hex/wedge–cooper scheme with 25 μm spacing in longitudinal direction. The volume for bottom substrate (heater) is subsequently created with hex/wedge–cooper scheme with 25 μm spacing in longitudinal direction tracing the surface meshes from main channel, oblique fins and oblique channels at the interface. A total of 2,398,560 hexahedral cells are generated for the computational domain.

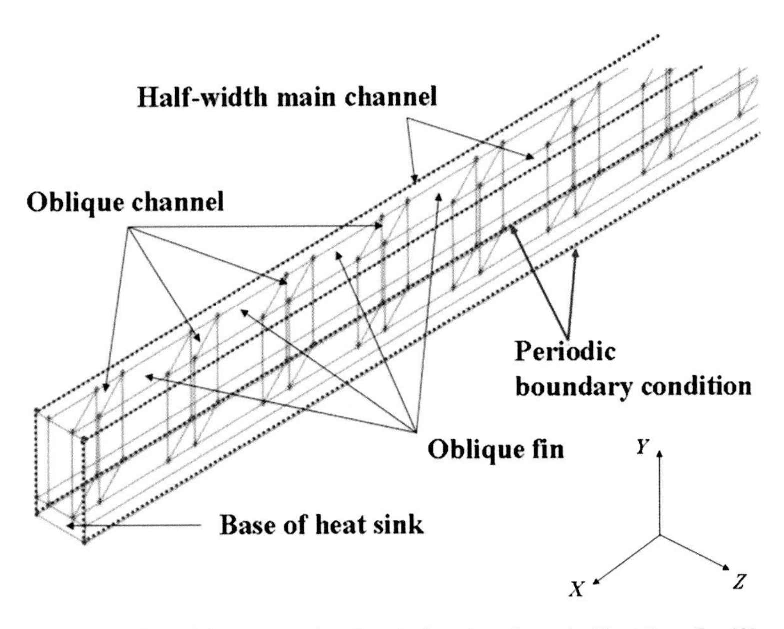

Fig. 2.5 Enlarged view of the computation domain for microchannel with oblique fins [3]

On the other hand, symmetry boundary condition is adopted at the centre of fin and channel for the conventional microchannel heat sink configuration, following the commonly used simplified boundary condition in the literature [8–10]. The symmetry boundary condition assumes that there is no fluid flow and heat transfer across the boundary surface. This effectively reduces the simulation domain to a half channel fin pair. The microchannel edges are then meshed by a 10 μm (spanwise) × 20 μm (longitudinal) × 50 μm (streamwise) spacing (50 × 90 × 500 cells). Both the fluid and solid regions are meshed with hex-map and hex-submap scheme, respectively, and a total of 2,250,000 hexahedral cells are generated.

The simulations are executed with the general purpose commercial CFD software, FLUENT v6.3, which solves the four governing equations numerically. These governing equations consist of continuity equation, momentum equation, energy equation for liquid and energy equation for solid listed as follows:

$$\nabla \cdot (\rho \vec{v}) = 0 \tag{2.1}$$

$$\nabla \cdot (\rho \vec{v} \vec{v}) = -\nabla P + \nabla \cdot (\mu \nabla \vec{v}) \tag{2.2}$$

$$\nabla \cdot (\rho \vec{v} c_p T) = \nabla \cdot (k \nabla T) \tag{2.3}$$

$$\nabla \cdot (k \nabla T) = 0 \tag{2.4}$$

Upon exporting the mesh files to FLUENT, the 3D double-precision pressure-based solver is selected with standard SIMPLE algorithm as its pressure–velocity coupling method. Standard discretization scheme is used for the pressure equation, while second-order upwind discretization scheme is selected for both momentum and energy equations. Water liquid (H_2O <l>) is chosen as the working fluid, while copper (Cu) with constant thermal conductivity, $k_{cu} = 387.6$ W/mK, is selected as fins and heat sink material. The density, specific heat capacity, thermal conductivity and dynamic viscosity of water are evaluated at the mean fluid temperature (average of the fluid inlet and outlet temperatures) using the following formulas [11]:

Density,

$$\rho(T) = \frac{a_0 + a_1 T + a_2 T^2 + a_3 T^3 + a_4 T^4 + a_5 T^5}{1 + bT} \tag{2.5}$$

where

$a_0 = 999.8396$	$a_4 = 1.49756 \times 10^{-7}$
$a_1 = 18.22494$	$a_5 = -3.93295 \times 10^{-10}$
$a_2 = -7.92221 \times 10^{-3}$	$b = 1.81597 \times 10^{-2}$
$a_3 = -5.54485 \times 10^{-5}$	T has the unit of °C

Specific heat capacity,

$$c_p(T) = 8,958.9 - 40.535T + 0.11243T^2 - 1.0138 \times 10^{-4} T^3 \tag{2.6}$$

Thermal conductivity,

$$k(T) = -0.58166 + 6.3555 \times 10^{-3} T - 7.9643 \times 10^{-6} T^2 \tag{2.7}$$

Dynamic viscosity,

$$\mu(T) = 2.414 \times 10^{-5} \times 10^{\left(\frac{247.8}{T-140}\right)} \tag{2.8}$$

For the calculation of specific heat capacity, thermal conductivity and dynamic viscosity, the temperature, T, should have unit of Kelvin (K). A residual of 10^{-6} is set as the convergence criterion for the continuity equation, x-velocity, y-velocity and z-velocity, while that for the energy equation is set as 10^{-9}.

The global coordinate system is defined such that X is in the axial direction (channel length), Y is in the longitudinal direction (channel depth) and Z is in the transverse direction (channel width). The local heat flux and local temperature distributions can be obtained from numerical simulations. With these quantities, the local convective coefficient, $h(X)$, can be calculated using the following equation:

$$h(X) = \frac{1}{A(X)} \frac{q(X)}{T_w(X) - T_f(X)} \tag{2.9}$$

where $A(X)$ and $q(X)$ are the total local heat transfer area and total local heat input, respectively, defined as follows:

$$A(X) = \sum_{Y,Z} dA(X,Y,Z) \tag{2.10}$$

$$q(X) = \sum_{Y,Z} q''(X,Y,Z)\, dA(X,Y,Z) \tag{2.11}$$

$T_w(X)$ and $T_f(X)$ on the other hand are local wall temperature and local fluid bulk-mean temperature given by

$$T_w(X) = \frac{\sum_{Y,Z} T_w(X,Y,Z)\, dA(X,Y,Z)}{\sum_{Y,Z} dA(X,Y,Z)} \tag{2.12}$$

$$T_f(X) = T_{f,\text{in}} + \frac{1}{\dot{m}C_p} \sum_{X,Y,Z} q''(X,Y,Z)\, dA(X,Y,Z) \tag{2.13}$$

As for the enhanced microchannel heat sink, it is difficult to compute for the actual local heat transfer coefficient due to the discontinuity between fins. Thus, the computational domain is axially divided into small control volumes, and the heat transfer coefficient is computed as a local average. Adopting the formulas mentioned above, $A(X)$, $q(X)$ and $T_w(X)$ are the total heat transfer area, total heat input and area-weighted average channel wall temperature of the control volume. Fluid bulk-mean temperature of the control volume is calculated as

$$T_{f,ob}(X) = \frac{1}{2}\left(T_{f,\mathrm{in}} + T_{f,\mathrm{out}}\right) \tag{2.14}$$

where $T_{f,\mathrm{in}}$ and $T_{f,\mathrm{out}}$ are the bulk fluid temperature at the inlet and outlet of the control volume. The average Nusselt number for the microchannels can then be computed as the axially weighted average values of $Nu(X)$ by

$$Nu_{\mathrm{ave}} = \frac{\sum Nu(X)\,dL}{L} \tag{2.15}$$

For both cases considered, a fully developed velocity profile is assigned to the inlets following the finding by Lee et al. [12] that flow in microchannels is mostly thermally developing. An average velocity of 0.3 m/s is set at the entrance of the microchannel, which corresponds to 13.5 mL/min volumetric flow rate through one single channel, while the total volumetric flow rate through the heat sink is 311 mL/min. The resultant Reynolds number, Re, in the main channels (conventional and enhanced microchannel) is at 273. Pressure outlet boundary condition is assigned to the outlets, where the flow is assumed to reach atmospheric pressure at the outlet of the microchannels. A uniform heat flux of 65 W/cm^2 is supplied from the bottom of the substrates, while the top surface of the copper microchannel is assumed bonded with adiabatic cover for sealing purposes.

The meshes used in both the simulation studies for conventional and enhanced microchannel heat sinks are verified to result in grid-independent results. The resultant average Nusselt numbers from different meshes used are in close proximity to each other. For instance, average Nusselt numbers of 6.89, 6.88 and 6.88 were obtained with the mesh counts of $34 \times 60 \times 333$ cells, $50 \times 90 \times 500$ cells and $66 \times 120 \times 625$ cells, respectively, for the case of conventional microchannel. The average Nusselt number varies by 0.15 % from the first to the second mesh and only by 0.08 % upon further refinement to the finest grid. Thus, the intermediate grid ($50 \times 90 \times 500$ cells) was selected. On the other hand, average Nusselt numbers of 12.25, 11.93 and 11.84 were achieved with the mesh count of $25 \times 45 \times 625$ cells, $40 \times 72 \times 1{,}000$ cells and $50 \times 90 \times 1{,}250$ cells, respectively, for the case of enhanced microchannel. The variations in average Nusselt numbers are 2.61 % from the first to the second mesh and 0.75 % from second to the finest grid. Likewise, the intermediate grid ($40 \times 72 \times 1{,}000$ cells) was selected for enhanced microchannel.

2.2.2 Fluid Flow and Heat Transfer Characteristics

Simulations showed clear distinctions between the flow fields of conventional and enhanced microchannel. Figure 2.6a illustrates the mid-depth plane ($Y=0.00075$ m, $Y'=0.5$) velocity profile of water in the microchannels at $X=0.0125$ m ($X'=0.5$). Conventional microchannel shows consistent velocity contour throughout the channel with high velocity gradient from the channel wall to the fluid core. The consistent and parabolic velocity contour suggests that the hydrodynamic boundary layer is fully developed and merged at the centre of the channel.

On the other hand, the breakage of the continuous fin into sectional oblique fins disrupts the velocity profile and thus the hydrodynamic boundary layer development at the trailing edge of each oblique fin section. The discontinuity with the downstream fin causes the hydrodynamic boundary layer development to restart at the leading edge of the next downstream fin. In addition, the oblique fin is much shorter, which limits the development of boundary layer when compared to the long continuous fin of the conventional microchannel. Thus, the velocity profile is renewed each time the coolant flows passes an oblique fin. In addition, the introduction of oblique channels diverts a fraction of coolant from the main channel into it.

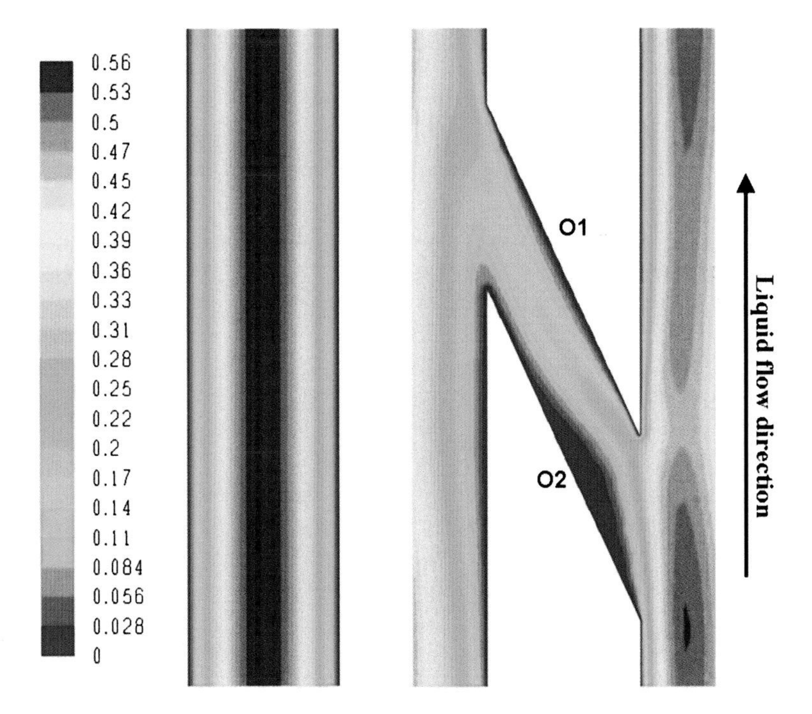

Fig. 2.6 Velocity contour (in m/s) of flow inside (**a**) conventional microchannel; (**b**) enhanced microchannel heat sinks at $X'=0.5$ and $Y'=0.5$ [3]

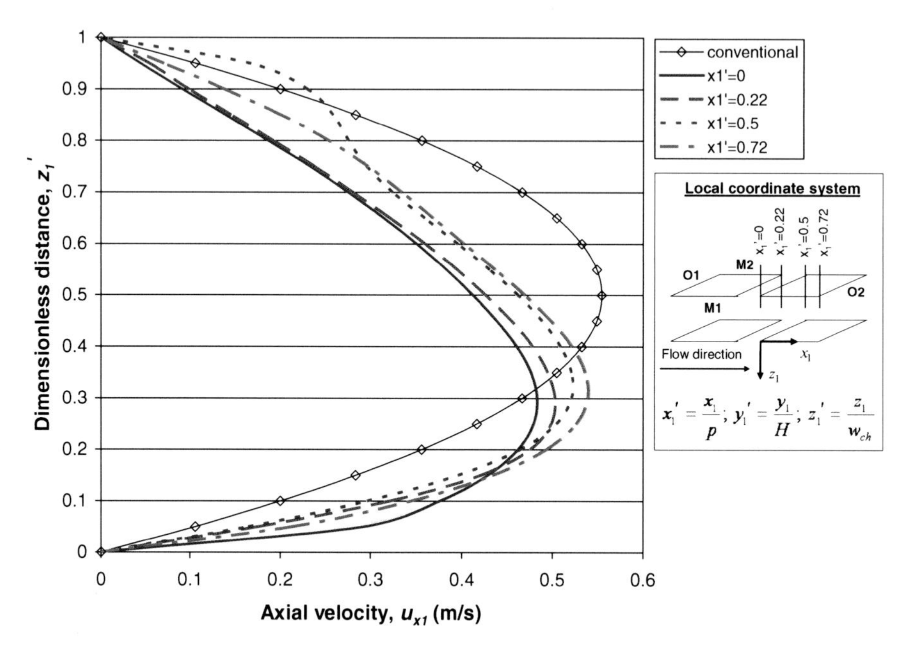

Fig. 2.7 Axial velocity profiles (in m/s) at the mid-depth plane of microchannel heat sinks at $X'=0.5$ and $Y'=0.5$ [3]

Owing to the axial flow momentum, the secondary flow created is brought closer to fin surface **O1**, as evident from the contour plot in Fig. 2.6b. The axial velocity profiles in Fig. 2.7, which are taken at the same position as in Fig. 2.6, reconfirm the thinning of boundary layer.

A local coordinate system is defined for the enhanced microchannel, originating from the tip of the oblique fin with x_1 lying along the main channel and z_1 in the transverse direction. Fin surfaces are also named specifically (M1, M2, O1 and O2). Uniquely skewed and asymmetrical velocity profile emerges in the main channel owing to the discontinuity between sectional fins and the asymmetry oblique channel arrangement. These axial velocity profiles of the enhanced microchannel skew towards fin surface M1 at $z_1'=0$ in the main channel compared to the fully developed velocity profile of conventional microchannel, which is symmetry at the centre line ($z_1'=0.5$). Consequently, the oblique fin surfaces M1 at $z_1'=0$ plane will have thinner boundary layer even though the average velocity in the main channel is lower compared to conventional microchannel. On the other hand, the oblique fin surfaces M2 on $z_1'=1$ plane have thicker hydrodynamic boundary layer.

As liquid coolant flows downstream to $x_1'=0.22$, hydrodynamic boundary layer grows thicker at $z_1'=0$ plane (fin surface M1), where velocity profile shifts upward towards the centre line, creating a more pointed profile. Between $x_1'=0.22$ and $x_1'=0.5$, the main channel receives additional liquid coolant (secondary flow) which is converging from the upstream oblique channel at $z_1'=1$. Thus, the coolant veocity

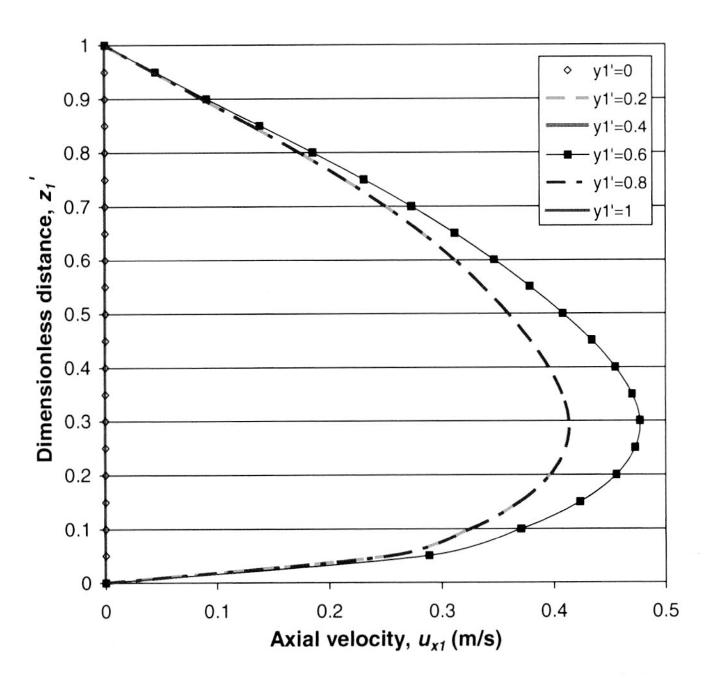

Fig. 2.8 Variation of velocity profiles (in m/s) in the longitudinal direction, y1′ [3]

in the main channel for $x_1'=0.5$ increases, especially in the region between $z_1'=0.8$ and $z_1'=1$, thinning the hydrodynamic boundary layer thickness at $z_1'=1$ plane (fin surface M2) significantly, while boundary layer continues to grow at $z_1'=0$ (fine surface M1). Subsequently, the coolant from oblique channel merges into the main channel flow smoothly leading to a skewed parabolic velocity profile as liquid coolant approaches $x_1'=0.72$. A glance at the variation of velocity profile along the longitudinal direction, y_1', in Fig. 2.8 shows that the velocity profiles are quite consistent where velocity is highest at the centre of the channel and slows down as it approaches the channel walls at $z_1'=0$, $z_1'=1$, $y_1'=0$ and $y_1'=1$.

Figure 2.9 continues to demonstrate the development of velocity profile along the fin. At $x_1'=0.72$, coolant flow in the main channel encounters a diverging oblique channel at $z_1'=0$. The presence of the smaller oblique channel induces a small fraction of the liquid in main channel into it and creates secondary flow. In this diagram, $z_1'>0$ denotes main channel region and $z_1'<0$ represents region in oblique channel. At $x_1'=0.88$, coolant flows into the diffusive oblique channel at $z_1'<0$ and shifts the velocity profile towards $z_1'<0$. As a result of secondary flow generation, coolant velocity in the main channel decreases while coolant velocity in the oblique channel increases. In addition, liquid coolant in the diffusive oblique channel encounters adverse pressure gradient due to the sharp increase in flow area. The flow near the oblique fin surface **O2** no longer has energy to move into the higher-pressure region that is imposed by the decrease in velocity at the edge of the boundary layer [13], causing slight boundary layer separation or flow reversal in oblique channel.

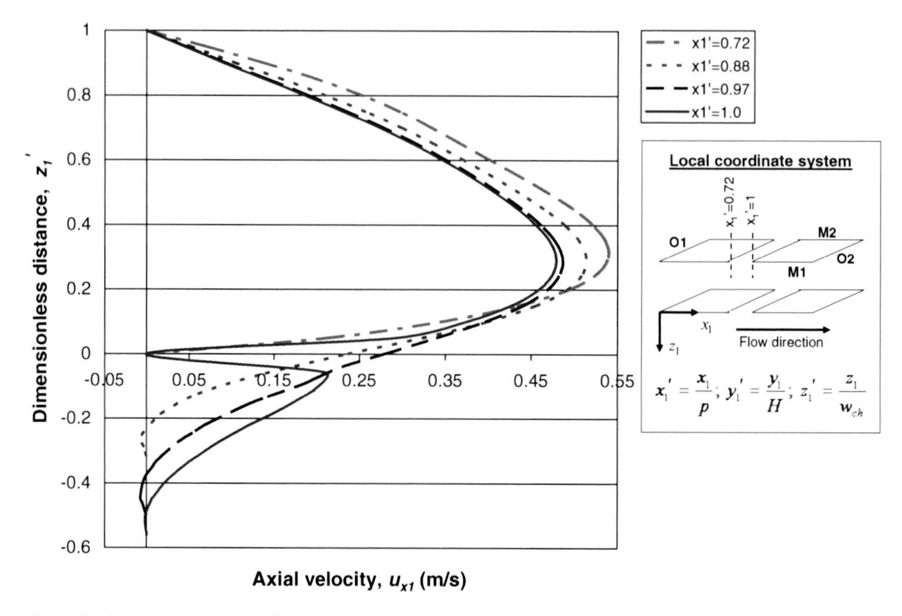

Fig. 2.9 Development of velocity profiles (in m/s) along the oblique fin [3]

Further downstream at $x_1'=0.97$, velocity profile assumes a relatively symmetry parabolic curve at $z_1'=0.3$ as coolant flows further into the oblique channel. Along with the secondary flow generation process, fluid particles with higher momentum are brought closer to fin surface M1 at $z_1'=0$ in the main channel. Finally as coolant reaches $x_1'=1$ (or $x_1'=0$ for downstream fin), shear stress halts the fluid particles that are in contact with the tip of downstream oblique fin at $z_1'=0$ (intersection of fin surface M1 and O1). However, liquid particles in the adjacent layers continue to flow with high velocity and create a skewed asymmetrical velocity profile with high near-wall velocity gradient in both main channel and oblique channel for fin surfaces M1 and O1 at $z_1'=0$. It is observed that the velocity profile at $x_1'=1$ (downstream of oblique fin) matches the velocity profile at $x_1'=0$ (upstream of oblique fin) very well, and this indicates that the velocity profile and boundary layer development is reinitiated as the coolant travels passed the oblique fins. Also, hydrodynamic boundary layer thickness at fin surface M1 at $z_1'=0$ is consistently thinner than that of the conventional microchannel heat sink throughout the streamwise positions examined. Although fin surface M2 at $z_1'=1$ displayed the opposite trend, its boundary layer thickness is significantly thinned down when surface M2 at $z_1'=1$ receives the secondary flow injection from oblique channel at $x_1'=0.5$. The effect from this series of velocity profile and boundary layer development on the overall heat transfer characteristic is discussed in the next section.

It is worth mentioning that the fluid flow characteristics of the current microchannel heat sink design are found to bear some similarities to that of fractal-shaped branching channel network. In the simulation results by Alharbi et al. [14], asymmetrical velocity profiles were present due to the asymmetric bifurcating angle in their fractal-shaped microchannel. Besides, hydrodynamic boundary layer redevelopment after a bifurcation and slight flow reversal in the bifurcations were also noticed by Alharbi et al. [14] and Senn and Poulikakos [15].

The unique oblique fin design modulates the flow and creates a constantly renewed and developing flow field that is highly desirable for heat transfer applications. Periodically reinitialized hydrodynamic boundary layers coupled with the short section of fins would lead to thinner hydrodynamic boundary layers, where similar development could be predicted for the thermal boundary layer. The renewed thermal boundary layer at each oblique fin section is thinner than that of the conventional continuous fin which thickens continuously from the inlet to the outlet. Besides generating secondary flows, the oblique cuts (channels) that are laid along the main channels increase the convective heat transfer surface area for the enhanced microchannel heat sink. This supplementary heat transfer area at the base (unfinned area) and fin area in the oblique channels provides an additional ~25 % heat transfer area in comparison with the conventional microchannel. The secondary flows that travel along the oblique channel would promote additional heat dissipation from the oblique surfaces of the fins.

Figure 2.10a displays the significant water temperature difference from the fluid core to the channel wall, ranging from 20 to 63 °C for the case of conventional microchannel.

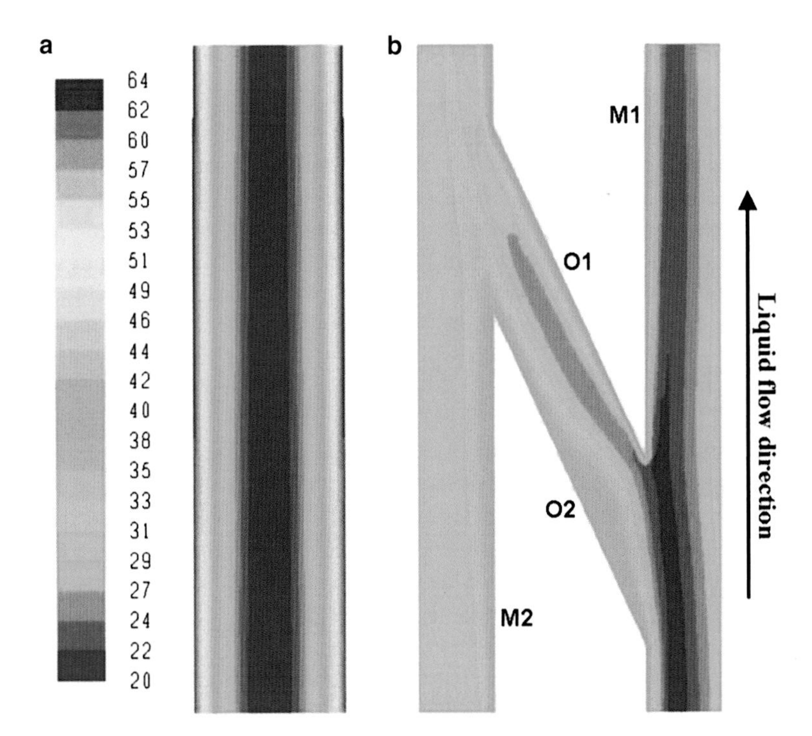

Fig. 2.10 Temperature contour (in °C) of flow inside (**a**) conventional microchannel; (**b**) enhanced microchannel heat sinks at $X'=0.5$ and $Y'=0.5$ [3]

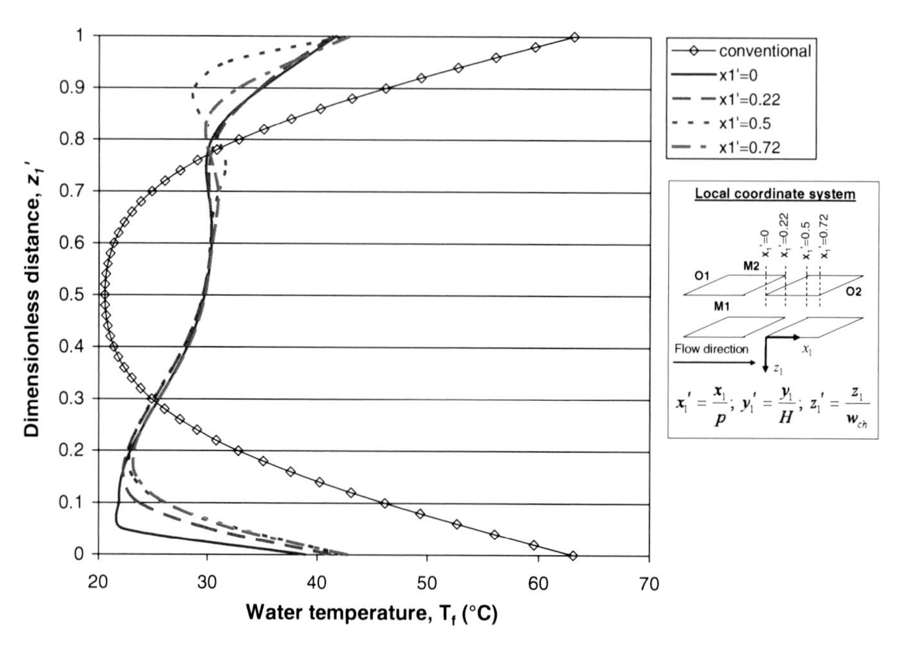

Fig. 2.11 Water temperature profile (in °C) at the mid-depth plane of microchannel heat sinks at $X'=0.5$ and $Y'=0.5$ [3]

As convection occurs, heat that is conducted from the channel wall into fluid particles at the wall propagates further into the fluid core through energy exchange with the adjoining fluid layer, causing the temperature gradients to develop in the fluid and thermal boundary layer to grow as coolant flows downstream. Continuous thickening of thermal boundary layer could lead to the deterioration of convective heat transfer and a further increase in wall temperature. In contrast, temperature contour of the coolant flow in the enhanced microchannel in Fig. 2.10b shows more uniform fluid temperature distribution in between 21 and 43 °C, as a result of better fluid mixing.

The water temperature profiles for both microchannel heat sinks are then compared in Fig. 2.11. Aligning with the skewed and asymmetrical velocity profile and the temperature contour showed previously, the temperature profiles of enhanced microchannel skew sharply to fin surface M1 at $z_1'=0$ planes. Comparing with the conventional microchannel, this fin surface M1 has much thinner thermal boundary layer with larger near-wall fluid temperature gradient. Thus, the combination of the entrance and secondary flow effects results in superior heat transfer performance, leading to lower wall temperature. Unlike Chandratilleke et al. [16] who employed high-frequency oscillating diaphragm that generates liquid jet to break up boundary layers and creates a steep temperature gradient, the current heat sink design achieved the similar feature through simple oblique fin design and does not require external power source.

Figure 2.11 also displays the development of temperature profiles along the fin. While the thermal boundary layer grows thicker along the fin with the near-wall fluid temperature gradients between water and channel walls are reducing, the injection of secondary flow into the main channel at $x_1'=0.5$ at $z_1'=1$ plane results in an opposite trend. The near-wall temperature gradient of fin surface M2 at $z_1'=1$ is observed to increase significantly as compared with the upstream position due to the coolant injection. Such thermal boundary layer thinning would improve the heat transfer performance for the fin surface M2 at the $z_1'=1$. As a result of secondary flow generation that promotes better fluid mixing, the water temperature profile of the enhanced microchannel is relatively flat compared with the parabolic curve of conventional microchannel. More importantly, the maximum fluid temperature and the range of fluid temperature in the enhanced microchannel are much smaller than those in the conventional microchannel, although both would share the same mass-weighted average temperature. For instance, water temperature adjacent to the fin surface of conventional microchannel can be as high as 63 °C while that of enhanced microchannel is drastically reduced to 43 °C. This demonstrates the importance of having well-mixed fluid, where the fin surface temperature can be lowered down substantially.

Figure 2.12 on the other hand shows the development in temperature profiles as coolant starts to diverge into oblique channel. The prominent finding from Fig. 2.12 is the emergence of the highly skewed temperature profiles for both fin surfaces M1 and O1 at $z_1'=0$ as coolant reaches the next fin tip, which is aligned with the observation in the velocity profiles. The renewal of temperature profiles at $x_1'=1$, recreating the significantly high temperature gradients or thinner thermal boundary layers

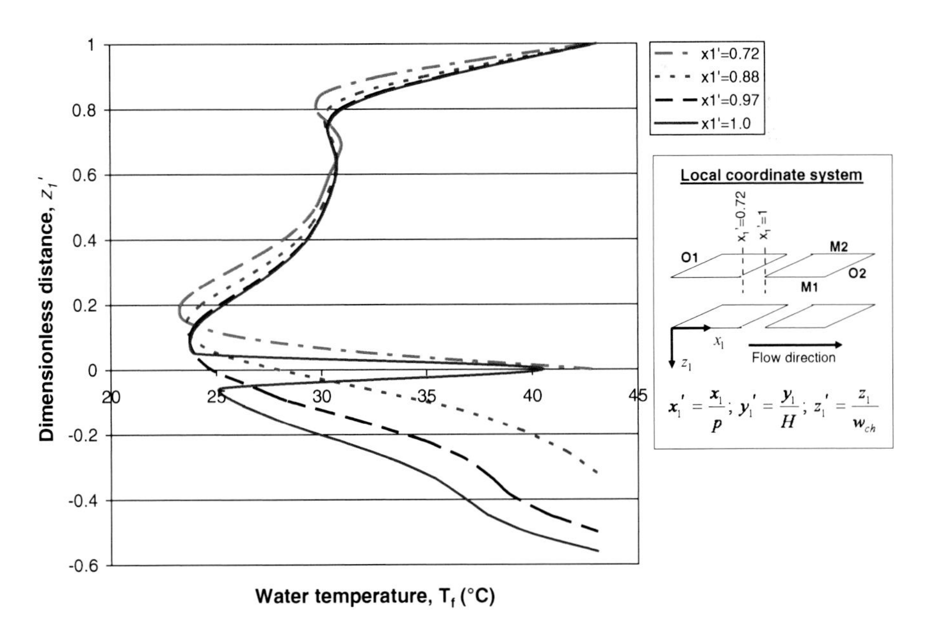

Fig. 2.12 Development of water temperature profile (in °C) along the oblique fin [3]

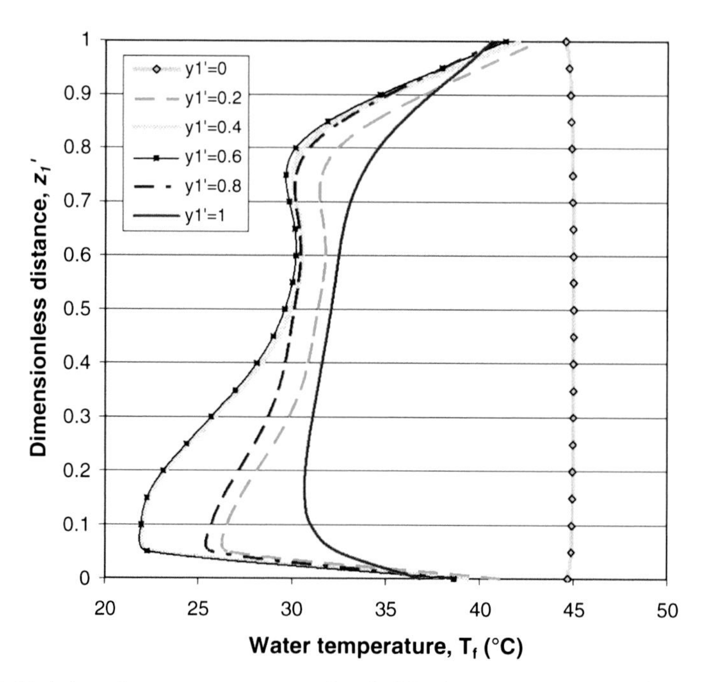

Fig. 2.13 Variation of water temperature profiles (in °C) with channel height at $X'=0.5$ [3]

where they are closely resembling the one at $x_1'=0$. Such phenomenon demonstrates the constantly renewal of thermal boundary layer at each oblique fin, keeping the thermal boundary layer thin for better heat transfer in comparison with conventional microchannel.

Variation of temperature profile with channel depth is then examined in Fig. 2.13. Except the bottom ($y_1'=0$) and top surface ($y_1'=1$) of microchannel, the other planes at various channel depth display a relatively consistent temperature profile among each other.

The combination of the boundary layers redevelopment and secondary flow effects results in superior heat transfer performance, leading to lower wall temperature compared to the conventional microchannel. Figure 2.14 shows that for the enhanced microchannel configuration, the maximum wall temperature is $T_{w,max}=53.2$ °C, while the temperature gradient is $\Delta T_w=13.7$ °C. The conventional microchannel heat sink, on the other hand, has a maximum wall temperature of 75.3 °C and a temperature gradient of 27.9 °C. Thus, the introduction of oblique fins resulted in the significant decrease of both the maximum wall temperature and temperature gradient of 22.1 and 14.2 °C, respectively.

In addition to the improvement in overall heat transfer performance, the presence of the oblique fins leads to significant local enhancement as illustrated in Fig. 2.15a. It is noticed that the initial heat transfer coefficient for conventional microchannel can be as high as ~15,000 W/m^{2}K. However, it diminishes quickly as the boundary layer thickens when the fluid travels downstream and attains a fairly constant value

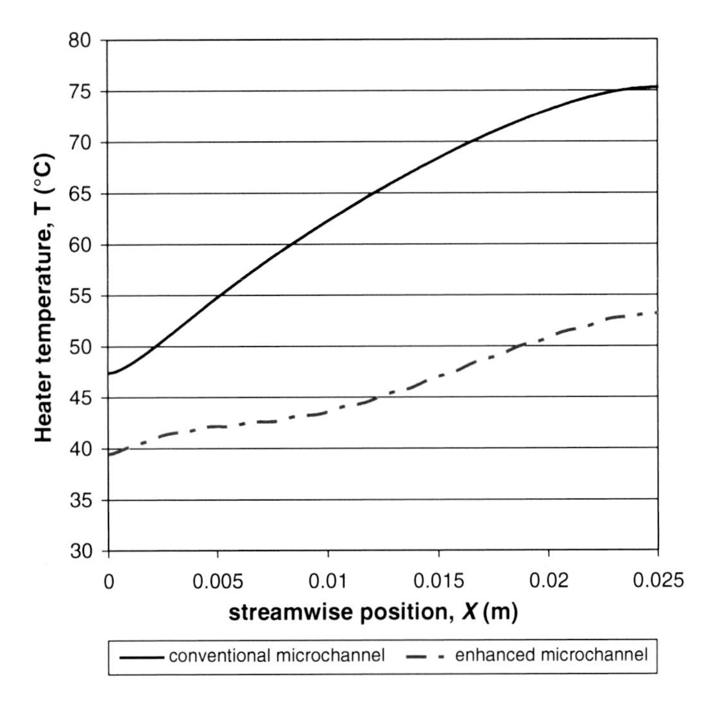

Fig. 2.14 Heater temperature profile for microchannel heat sinks [3]

at ~5,000 W/m^{2}K, displaying a highly non-uniform heat transfer performance from the inlet to the outlet of the heat sink. In contrast, the variation of local heat transfer coefficients for the enhanced microchannel is smaller, with a higher averaged value of ~10,000 W/m^{2}K. The local heat transfer coefficient is observed to increase 100 % almost everywhere. With the more effective heat transfer, 60 % more heat flux can be dissipated with the enhanced microchannel if the same maximum wall temperatures are maintained.

Moreover, it is observed that the level of heat transfer enhancement correlates closely with the percentage of secondary flows that are diverged through the oblique channel. Figure 2.15b illustrates the local heat transfer coefficient of the enhanced microchannel with the corresponding secondary flow percentage at each oblique channel branch. In this case, secondary flow rate percentage is defined as the proportion of coolant mass flux into oblique channel over coolant mass flux across the channel inlet. At the upstream of the microchannel heat sink, secondary flow rates are low and there is not much disruption to the coolant flow. Thus, heat transfer coefficient drops drastically from 15,000 to 8,000 W/m^{2}K as thermal boundary layer develops. Subsequently, secondary flow that is diverted to flows into adjacent channel increases the transverse momentum of the coolant flow in the main channel and causes the secondary flow rate (or momentum) to build up and maintain at around 14 % for the downstream oblique channels. As a result, local heat transfer coefficient increases and hovers at a higher value at ~10,000 W/m^{2}K.

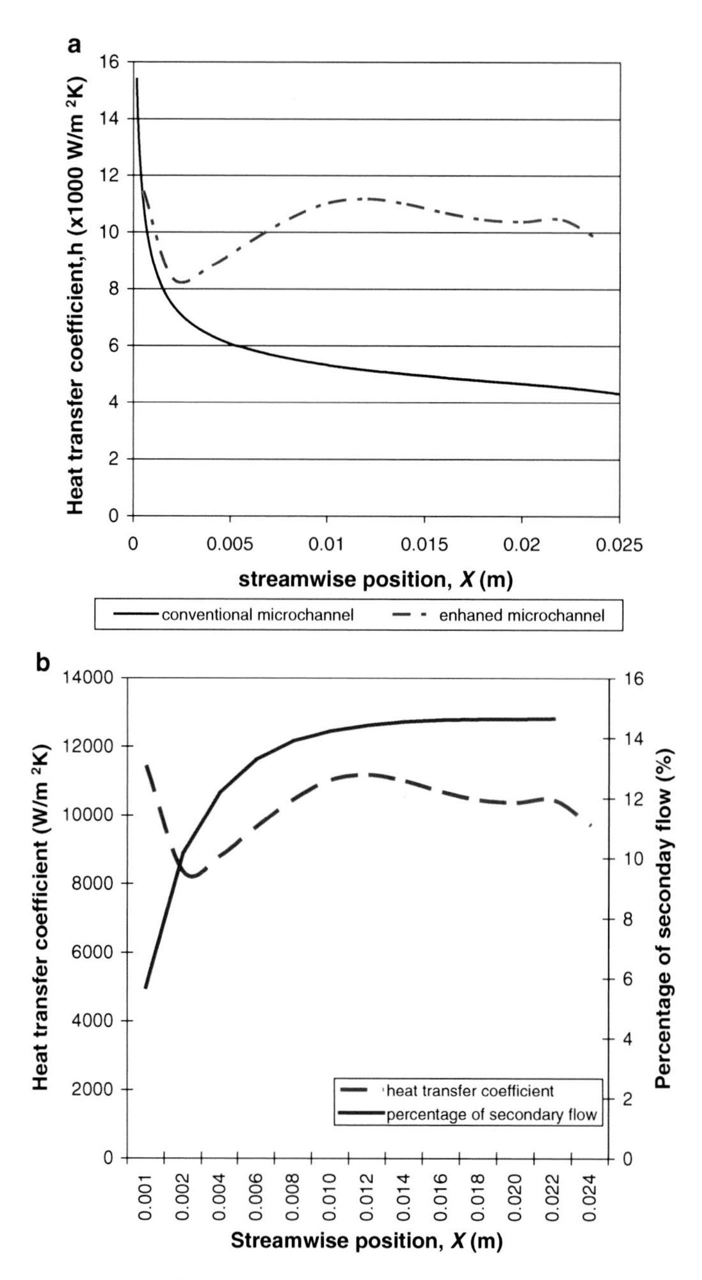

Fig. 2.15 Heat transfer coefficient profile for (**a**) comparison between microchannel heat sinks; (**b**) enhanced microchannel with percentage of secondary flows [3]

Steinke and Kandlikar [1] first proposed alternating the secondary channel direction in the streamwise direction with the intention to prevent flow migration. However, based on current simulation finding, it is highly possible that their configuration would not be able to generate sufficiently strong secondary flow as the transverse momentum in the flow is negated by the secondary flow in the opposite direction after each oblique channel branch. Sahnoun and Webb [17] developed the equations to predict flow efficiency in the louvred fin heat exchanger. In the context of louvred fin heat exchanger, flow efficiency is defined as the degree to which the flow follows the louvres, which is comparable to the percentage of secondary flow in the enhanced microchannel. Unfortunately, their equations cannot be adopted to predict the percentage of secondary flow for enhanced microchannel. The predicted secondary flow rate using the proposed equations by Sahnoun and Webb [17] is ~84 % compared to 14 % indicated in the simulation. The disparity is believed to be caused by the difference in the fin geometry in which oblique fin is much thicker than louvre fin. The thicker oblique fin reduces the oblique channel width significantly and results in much lower secondary flow percentage.

In comparison with enhanced microchannel, conventional microchannel experiences severe axial conduction or heat flux redistribution as a result of the non-uniform convective heat transfer performance as evident from Fig. 2.16. It is observed that channel walls at the inlet section dissipate 80 % more heat flux than the rest of the convective surfaces. Enhanced microchannel, on the other hand, dissipates 30 % more heat flux at the inlet region than the rest of heat transfer surfaces.

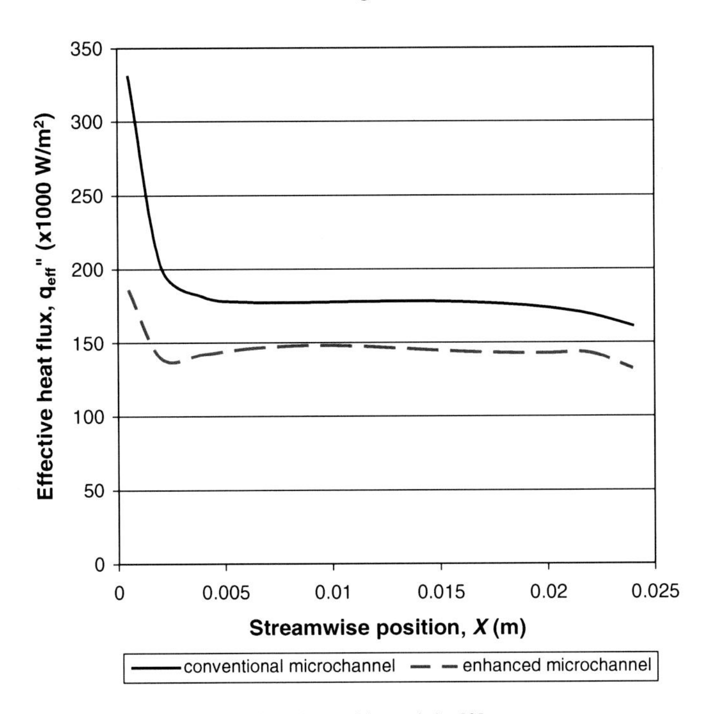

Fig. 2.16 Effective heat flux along microchannel heat sinks [3]

Such phenomenon leads to significantly higher temperature at the inlet region (as seen in Fig. 2.14) even though conventional microchannel exhibits comparable heat transfer coefficient as enhanced microchannel at the inlet. Owing to the larger heat transfer area, the average effective heat flux of enhanced microchannel is found lower than the one of convention microchannel.

Owing to the unique hydrodynamic and thermal boundary layer profiles in the enhanced microchannel heat sink, the amount of heat removed through each fin surface and the un-finned surface differs significantly, as shown in Fig. 2.17a.

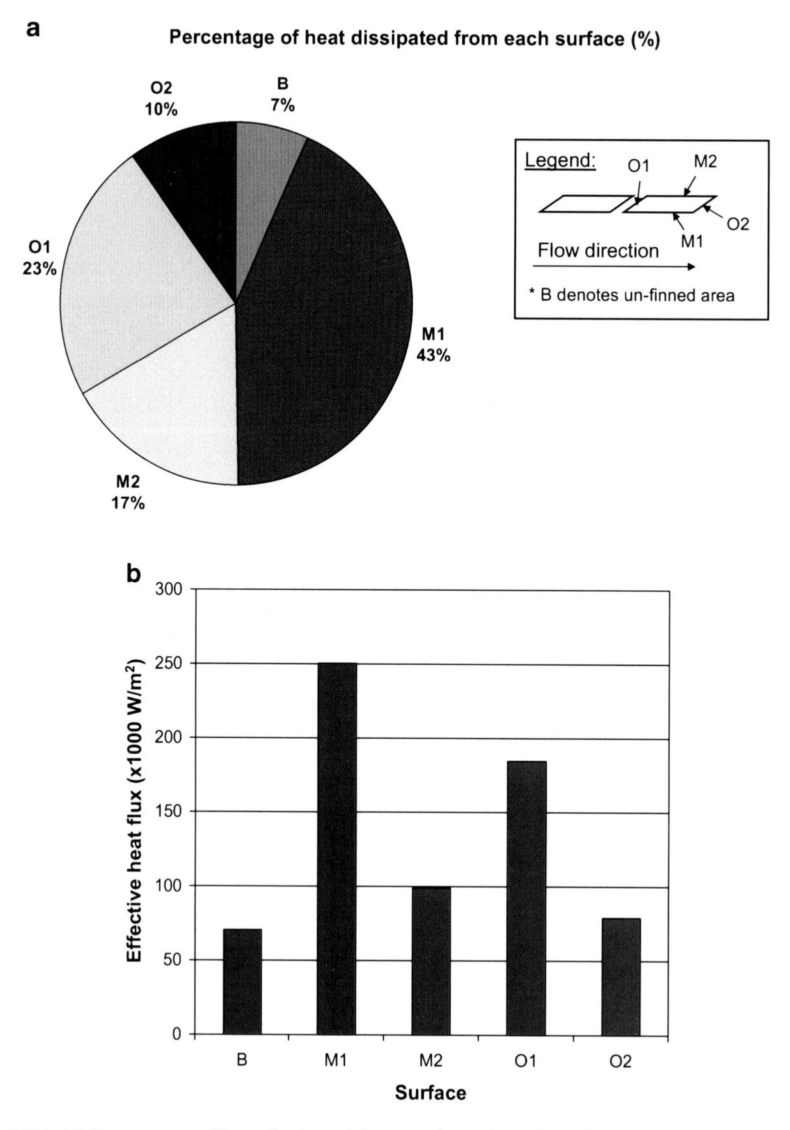

Fig. 2.17 (**a**) Percentage of heat dissipated from each surface; (**b**) effective heat flux dissipated from each surface [3]

Fin surfaces M1 and O1 record the highest percentage of heat removal with 43 and 23 % of total heat dissipated from each surface, respectively. Interestingly, surface M2, which is located alongside with surface M1 in the main channel (with equivalent heat transfer area and coolant mass flux), does not perform as good as surface M1. The fin surface M1 dissipates ~60 % more heat flux than surface M2. Moreover, surface O1 which has smaller heat transfer area and coolant mass flux (located in oblique channel) also removes heat more effectively compared with surface M2 (heat transfer area, M2: 25 %, O1: 18 %; average coolant mass flux, M2: 87 %, O1: 13 %). This phenomenon is mainly contributed by the skewed hydrodynamic and thermal boundary layer profiles as observed in Figs. 2.9 and 2.12. Surfaces M1 and O1 have much thinner boundary layer thickness compared with the rest of the fin surface and un-finned surface, thus managing to achieve highly augmented heat transfer. Figure 2.17b on the other hand displays the effective heat flux that is transferred through each surface, where surfaces M1 and O1 remove 250 and 190 kW/m^2, respectively, while the rest of the surfaces dissipate less than 100 kW/m^2. This proves that surface M1 and O1, which are associated with thinner boundary layers than the rest of the surfaces, are the key surfaces in this passive heat transfer enhancement technique.

This novel idea for enhancing the heat transfer performance of microchannel heat sinks is attractive as the pressure drop penalty is small. As shown in Fig. 2.18, the pressure drop for the enhanced microchannel heat sink with oblique fins is only ~10 % higher than that of a conventional microchannel heat sink. For the case of conventional microchannel, the pressure profile decreases linearly as the flow is

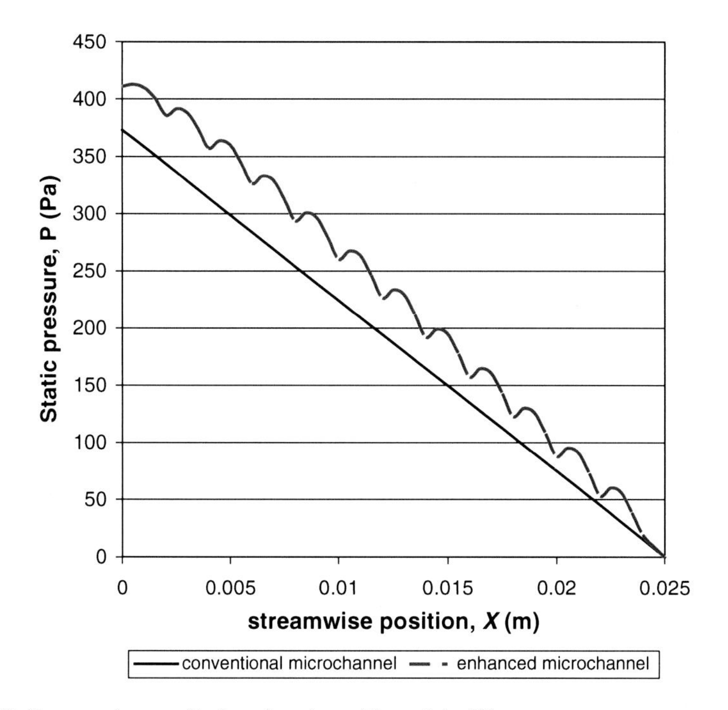

Fig. 2.18 Pressure drop profile for microchannel heat sinks [3]

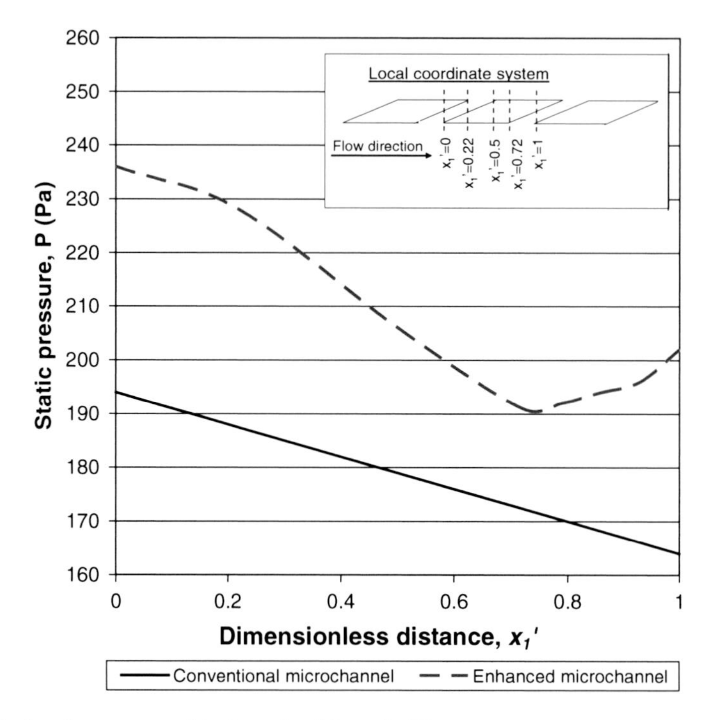

Fig. 2.19 Local pressure profile along the oblique fin [3]

fully developed hydrodynamically. On the other hand, the pressure profile for enhanced microchannel is characterized by local drops and recoveries.

Exploring further into the local pressure drop profile along the fin in Fig. 2.19 shows that pressure drop occurs for $x_1' < 0.72$ while pressure rise occurs for $x_1' > 0.72$. It is also noticed that the local pressure drop for the enhanced microchannel heat sink is steeper than that for the conventional microchannel. This could be due to the periodic redevelopment of hydrodynamic boundary layers which results in the thinning of the boundary layers. The local pressure drop is immediately followed by local pressure recovery when the flow from the main channel branches into the oblique channel.

This local pressure recovery was also observed by Alharbi et al. [14] in their numerical study on fluid flow through microscale fractal-like branching channel networks. They argued that the observed pressure recovery at each bifurcation may be due to the tapered increase in cross-sectional area similar to a "diffuser" following each bifurcation. The pressure recoveries compensate for the increased pressure drops, resulting in only a small overall pressure drop penalty. Similarly, Xu et al. [18, 19] observed the pressure recovery effect as coolant expands into the re-entrant chamber. The proposed oblique fin layout is different from the conventional passive heat transfer enhancement techniques, where there is a hefty trade-off in terms of pressure drop penalty that is always associated with the improved heat transfer.

For instance, pin-fin heat sinks achieved improved thermal performance over plate-fin heat sinks by hindering the development of the thermal boundary layer in a unidirectional flow at the expense of a significant pressure drop penalty [20].

2.3 Numerical Analysis: Full Domain Simulation

Numerical analysis employing full domain heat sink model is conducted to investigate the flow migration phenomenon and its effect on the overall fluid flow and heat transfer characteristics in the enhanced microchannel heat sink. A unique coolant distribution exists in the main channels and oblique channels as a result of continuous secondary flow generation and coolant migration. Simulation findings are in close proximity to that of the simplified numerical model for low flow rate case where both models display comparable fluid flow (velocity profiles) and heat transfer characteristics. However, it is also found that the simplified model, which assumes infinite span of oblique fins, over-predicts the heat transfer performance when the Reynolds number is higher.

2.3.1 Simulation Model

This full domain microchannel heat sink is modelled based on the actual silicon-based test vehicle used in experimental investigation [21] and incorporates inlet and outlet manifolds. Detailed geometries and the mesh intervals for the model are tabulated in Table 2.2, while Fig. 2.20 displays the sketch of oblique fin microchannel heat sink using oblique fins, where O1, O2, M1 and M2 are the different surfaces of oblique fin [22].

Table 2.2 Dimensional details of oblique fin microchannel heat sink model

Characteristic	Oblique fin microchannel
Material	Silicon
Heat sink footprint, width × length (mm)	12.7 × 12.7
Number of fin row, N	61
Main channel width, w_c (μm)	113
Fin width, w_w (μm)	87
Channel depth, H (μm)	379
Aspect ratio, α	3.35
Oblique channel width, w_{ob} (μm)	49
Fin pitch, p (μm)	405
Fin length, l (μm)	292
Oblique angel, θ (°)	26.3
Mesh interval x, y, z (simplified model)	1,270 × 65 × 40
Mesh interval x, y, z (full domain model)	1,393 × 33 × 1,059

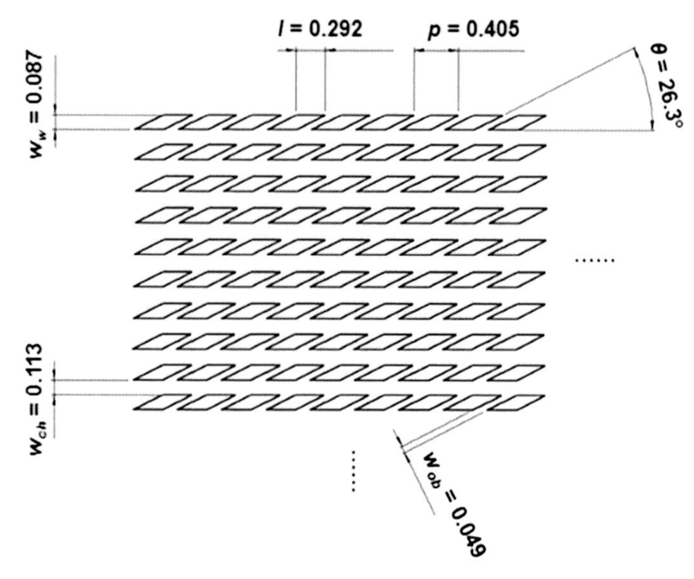

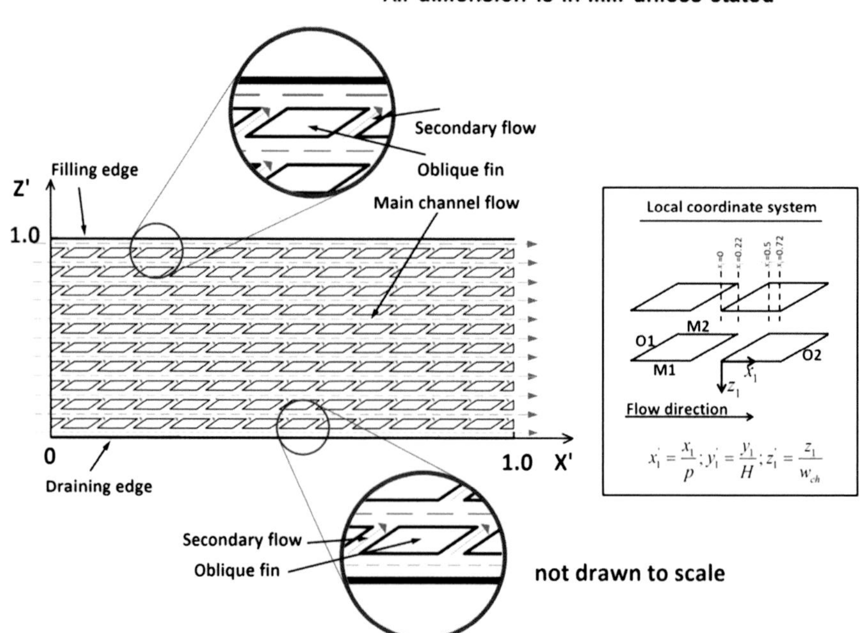

Fig. 2.20 Sketch of oblique fin microchannel heat sink

Emulating the silicon thermal test chips used in the experimental studies, the 12.7×12.7 mm^2 microchannel heat sink model is divided into a cluster of 25 thermal test chips in a 5×5 grid array. Each thermal test chip is 2.54×2.54 mm^2 with a heater measured at 2×2 mm^2 located at the centre of silicon chip active surface.

Transient solver is adopted for full domain simulation as a result of the consideration of flow instability. The nature of flow in the oblique fin microchannel heat sink could be unsteady or periodic due to secondary flow generation and flow migration. The four governing equations solved in FLUENT v6.3 under transient solver consist of continuity equation, momentum equation, energy equation for liquid and energy equation for solid listed as follows:

$$\frac{\partial \rho}{\partial t} + \nabla \cdot \left(\rho \vec{v} \right) = 0 \tag{2.16}$$

$$\frac{\partial}{\partial t}\left(\rho \vec{v} \right) + \nabla \cdot \left(\rho \vec{v} \vec{v} \right) = -\nabla P + \nabla \cdot \left(\mu \nabla \vec{v} \right) \tag{2.17}$$

$$\frac{\partial}{\partial t}\left(\rho \vec{v} c_p T \right) + \nabla \cdot \left(\rho \vec{v} c_p T \right) = \nabla \cdot \left(k \nabla T \right) \tag{2.18}$$

$$\frac{\partial}{\partial t}\left(\rho c_p T \right) + \nabla \cdot \left(k \nabla T \right) = 0 \tag{2.19}$$

As for the solver configurations, the 3D double-precision pressure-based solver is selected with first-order implicit as its unsteady formulation. SIMPLE algorithm is set as the pressure–velocity coupling method. Standard discretization scheme is used for the pressure equation, while second-order upwind discretization scheme is selected for both momentum and energy equations. Water liquid (H_2O <l>) with its physical properties evaluated at mean temperature is chosen as the working fluid, while silicon (Si) with constant thermal conductivity, k_{si} = 148 W/mK, is selected as fins and heat sink material. A residual of 10^{-6} is set as the convergence criteria for the continuity equation, *x*-velocity, *y*-velocity and *z*-velocity, while that for the energy equation is set as 10^{-9}.

Different flow rates are set for the full domain model as compared with the data from experimental studies [21], while the analysis and comparison mainly focus on low flow rate setting, Re = 250, and high flow rate setting, Re = 660. For all cases considered, a uniform inlet pressure is assigned to the inlet. Inlet pressure is adjusted to achieve the targeted mass flow rate or Reynolds number. Furthermore, pressure outlet boundary condition is assigned to the outlet, where the flow is assumed to reach atmospheric pressure at the outlet of the heat sink. Each of the 25 heaters is supplied with heat flux at 274 W/cm^2 while the top surface of the silicon-based microchannel is assumed bonded with adiabatic cover for sealing purposes. Courant numbers for the low flow rate and high flow rate cases are 16.7 and 5.2, respectively. As the full domain simulation model adopts implicit unsteady formulation, a larger Courant number can be used to achieve fast convergence.

The current mesh count for full domain model of 1,393 × 33 × 1,059 results in a total of 42,915,914 cells. The mesh interval for the full domain model is replicated into the simplified model to study the grid sensitivity [3]. The resultant mesh count in the simplified model is 1,058 × 33 × 17, and the computed average Nusselt number is 18.48 when Re = 660. The variations in average Nusselt numbers are 3.5 % from the mesh density for full domain model to the mesh interval selected for the simplified model. Although the mesh density is reduced to accommodate for the large full domain model simulation, the selected mesh interval still provides reasonable accuracy in prediction. In addition, the simulation results for the full domain model would be verified against the experimental findings for its validity.

2.3.2 *Validation of Numerical Simulation*

Figure 2.21 shows the total pressure drop and average Nusselt numbers comparisons with different Reynolds numbers for the oblique fin microchannel heat sink between results from full domain simulation model and experimental data [22]. It is found that the maximum deviation of total pressure drop between numerical and experimental results is less than 7 % under all Reynolds numbers. Unfortunately, the computed average Nusselt numbers for oblique fin microchannel after temperature correction are about 20 % lower than the predictions by full domain simulation model. This could be caused by the lack of information in the measurability of thermal sensors leading to an inaccurate temperature correction. A trial using −2.7 °C as the correction factor in computation results in Nusselt numbers that are closely imitating the simulation results. This indicates that the numerical simulation studies are validated by the experimental measurement.

2.3.3 *Fluid Flow and Heat Transfer Characteristic*

The main objective of the full domain simulation model is to scrutinize the coolant mass distribution and flow instability for oblique fin microchannel heat sink and its effect to fluid flow and heat transfer. In this section, the simulation results for the full domain model are discussed when Re is 250 and 660.

Figure 2.22 shows the flow distribution of coolant in every main channel along the streamwise direction when Re = 250. At the upstream ($X'=0.01$) of the oblique fin microchannel heat sink, significant variation is observed for the mass flow rate in the main channel along the dimensionless heat sink width, Z'. Apart from the region between $0.4<Z'<0.8$, which has a comparable mass flow rate at 4.1×10^{-5} kg/s, the mass flow rate is significantly higher in the main channels at $Z'<0.4$ and lower in the main channels at $Z'>0.8$. The highest mass flow rate occurs in the first main channel in the spanwise direction ($Z'=0.02$) at 5.9×10^{-5} kg/s, which is almost 44 % higher than the nominal mass flow rate of the main channels. On the other hand, the lowest mass flow rate occurs in the last main channel in the spanwise direction ($Z'=0.98$) at 2.9×10^{-5} kg/s, a 29 % off from the nominal main channel mass flow rate. This indicates that the coolant mass distribution is skewed towards $Z'<0.5$ in the heat sink for this upstream position. This phenomenon occurs due to the fact that the coolant mass is forced to redistribute in the inlet manifold in order to maintain a

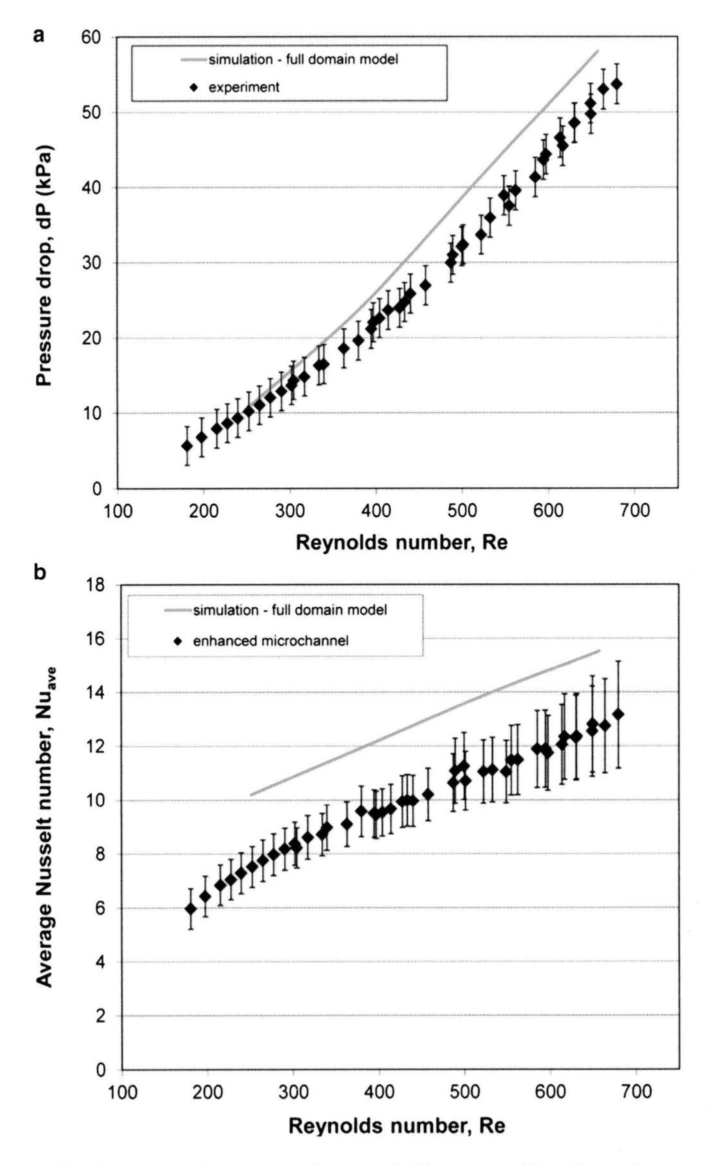

Fig. 2.21 (**a**) Total pressure drop comparisons and (**b**) average Nusselt number comparisons between numerical results and experimental data for the oblique fin microchannel heat sink

uniform pressure between the inlet and outlet as flow migration takes place within the heat sink. It is also noticed that except for $Z' > 0.8$, the mass flow rate in the main channel at $X' = 0.01$ is at least 10 % higher than that of the downstream locations. Low secondary flow rates in the oblique channels at $X' = 0.01$ is identified as the contributing factor and this will be elaborated further in the later section.

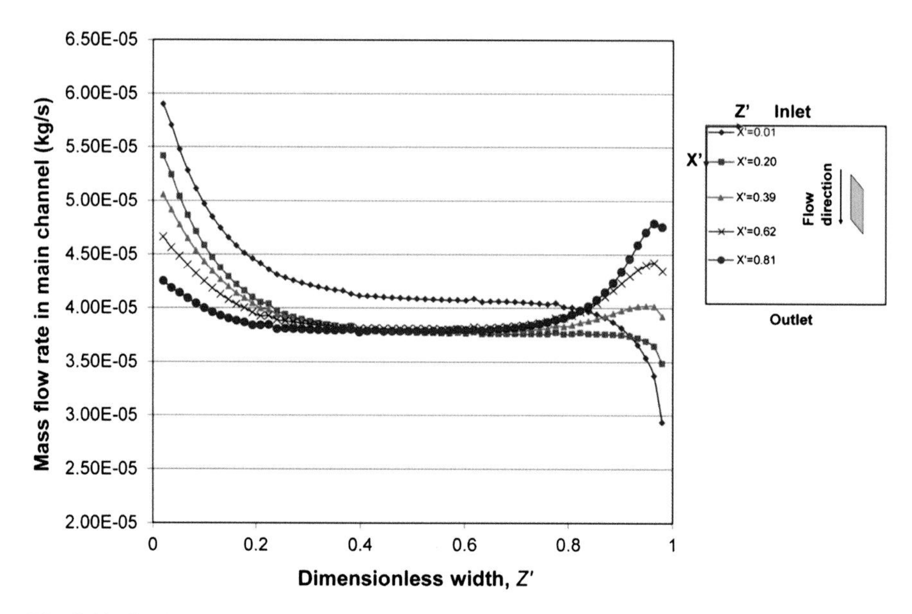

Fig. 2.22 Coolant mass flow rate distribution in the main channels of oblique fin microchannel heat sink when Re = 250

As coolant travels downstream, secondary flow generation and flow migration occur continuously. Mass flow rate in the main channels for $Z' < 0.3$ keeps reducing, while mass flow rate in the main channels for $Z' > 0.7$ rises consistently as it is illustrated in Fig. 2.22. This trend clearly demonstrates the coolant migration phenomenon. Coolant mass is observed to migrate from $Z' < 0.5$ half of the heat sink to $Z' > 0.5$ half of the heat sink as coolant flows downstream. Thus, small regions at $Z' > 0.8$ and $X' < 0.2$ with low coolant mass exist as a result of the coolant migration in the diagonal direction following the direction of secondary flow. Despite all the flow migration, the main channels located between $Z' = 0.25$ and $Z' = 0.75$ maintain a relatively consistent mass flow rate at all streamwise locations examined. This area constitutes 50 % of the heat sink and is influenced by the flow migration.

Examining Fig. 2.23 provides some insights to the coolant distribution in the oblique channels within the heat sink. Generally, the coolant flow rate in the oblique channels is an order of magnitude lower than that of the main channels. These oblique channels are sized such that they generate sufficiently strong secondary flows or momentums to disrupt the boundary layer development and promote fluid mixing, while avoiding unnecessary disturbance to the flow. The mass flow rate through the oblique channels at the upstream of the heat sink ($X' = 0.01$) is significantly lower compared with other downstream locations. When the coolant first enters the oblique fin microchannel heat sink, it flows primarily in the streamwise direction and has a very limited transverse velocity component or momentum. Oblique channels on the other hand branch off from the main channels with their flow path swayed away from the streamwise direction. Thus, the fraction of coolant

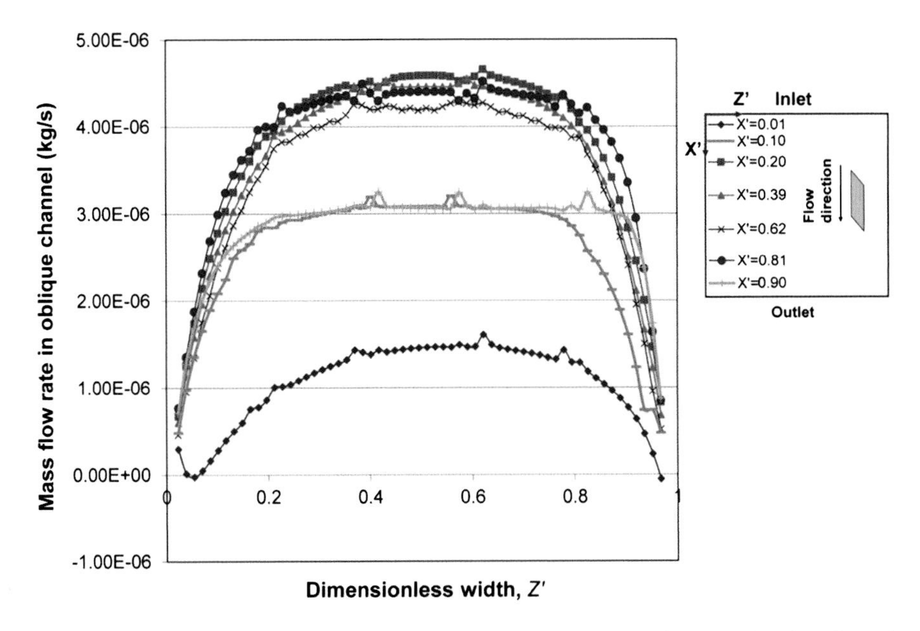

Fig. 2.23 Coolant mass flow rate distribution in the oblique channels of oblique fin microchannel heat sink when Re = 250

in the main channels that diverges into the oblique channels is limited at this upstream position. However, as the coolant flows downstream, the transverse velocity component and momentum build up as flow continues to divert into oblique channels, before being injected to the adjacent main channels. Consequently, secondary flow rates through the oblique channels elevate to a higher value (about 10 % of total coolant mass flow rate except in edge regions) and remain consistent throughout the heat sink from $X' = 0.20$ onwards. There is fluctuation in the secondary flows at different downstream locations, but the magnitude of the variation is not significant.

An "n" shape parabolic curve also indicates that secondary flow rates in the oblique channels located closer to the heat sink edge ($Z' < 0.2$ and $Z' > 0.8$) are lower than that of the oblique channels at the centre of the heat sink. Similar phenomenon is also observed by Dejong and Jacobi [2] in their flow visualization study on louvred fin heat exchanger. They concluded that wall effect existed on several fin rows next to the walls, where flow is characterized by lower flow efficiency (low secondary flow rate) and lower heat transfer. Low secondary flow rates in the oblique channels at the edge of the heat sink are also believed to result in lesser heat transfer enhancement compared with those at the centre of the heat sink in the context of enhanced microchannel. However, there is still at least 60 % of the total heat sink width having consistent secondary flow rates, and it will be explored further in the later sections whether this affects the heat transfer performance of the heat sink globally and locally.

Besides scrutinizing the macroscopic effect of the flow migration to the overall flow distribution, the full domain simulation highlights the development of local coolant velocity and temperature profiles at different spanwise and streamwise positions. Figure 2.24a displays the coolant velocity profiles at the mid-depth plane in multiple main channels at $X'=0.01$ ($x_1'=0$). Generally, developing velocity profiles emerge in the main channels at this upstream position of the heat sink just as

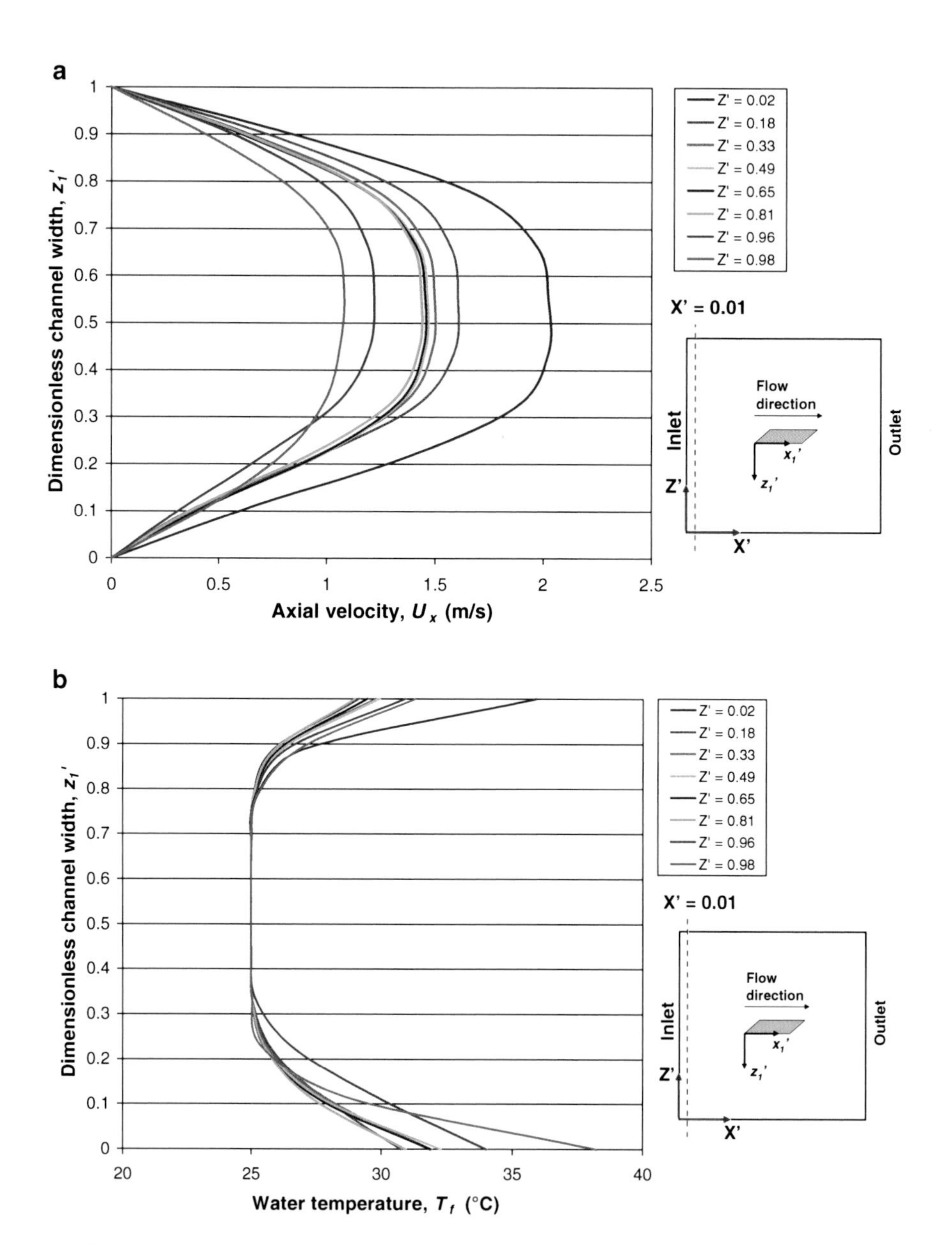

Fig. 2.24 Coolant (**a**) axial velocity and (**b**) temperature profiles in the main channels at $X'=0.01$, $Y'=0.5$, when Re = 250

coolant enters the heat sink. Velocity magnitudes are observed to reduce as Z' increases, aligning well with the observation from Fig. 2.3, where coolant mass is concentrated towards $Z'<0.5$ half of the heat sink. As a result, velocity profiles skew slightly towards $z_1'=1$ in the main channels. However, the velocity profiles are almost symmetrical at the centre of the channels, as coolant flows primarily in the axial direction. The resultant coolant temperature profiles at the mid-depth plane are showed in Fig. 2.24b. Similarly, developing temperature profiles are displayed as most of fluid temperature remains at the inlet temperature. Coolant temperature in the main channel at $Z'=0.98$ is significantly higher compared to the rest of the channels as this channel receives the least coolant mass among others.

As coolant travels downstream to $X'=0.20$, velocity profiles continue to develop and become significantly different from that of $X'=0.01$. Figure 2.25a displays the velocity profiles in the main channels where most of them skew towards $z_1'=0$ due to the divergence of fluid into oblique channels, which brings the fluid particles from the centre of the channel closer to fin surface **M1** ($z_1'=0$). As a result, plane $z_1'=0$ (fin surface **M1**) has the thinner boundary layer thickness compared to plane $z_1'=1$ (fin surface **M2**). Details and discussions about the boundary layer development along the sectional oblique fin have been studied by present authors in [22]. Coolant in the main channels between $0.18<Z'<0.81$ maintains a consistent velocity profile while the others display varying velocity profiles under the influent of flow migration and non-uniform secondary flow rate. On the other hand, the main channel at $Z'=0.02$ continues to show the highest axial velocity than the rest. Besides, the main channel at $Z'=0.02$ together with those main channels at $Z'>0.9$ display a velocity profile that is more symmetry than those between $0.18<Z'<0.81$ as secondary flow rates are lesser in these regions. Consequently, coolant temperature profiles at the mid-depth plane are very different from each other as demonstrated in Fig. 2.25b. Generally, skewed coolant temperature profiles emerge in the main channels between $0.18<Z'<0.81$ owing to the skewed velocity profiles that present in the channels. Water temperatures in this region are more uniform, and wall temperatures (temperature at $z_1'=0$ and $z_1'=1$) are also found to be lower than in the channels outside this region. This indicates that the fluid mixing and heat transfer from the channel surfaces between $0.18<Z'<0.81$ is more efficient than the rest. The effect of the skewed coolant water temperature profiles to the heat transfer was discussed in detail in the previous publication [22]. Wall temperature is the highest on surface $z_1'=1$ and $z_1'=0$ for main channels at $Z'=0.02$ and $Z'=0.98$, respectively, as these two channels are located at the edge of the heat sink, where the heat dissipation is concentrated onto these surfaces.

Moving forward to $X'=0.62$ in Fig. 2.26a, similar velocity profiles are observed for main channels within $Z'=0.18$ and $Z'=0.81$ with minimum change compared with those at the axial distance of $X'=0.20$. However, the axial velocity in the main channel at $Z'=0.02$ declines, while the axial velocity in the main channels at $0.96<Z<0.98$ rises as coolant continues to migrate towards $Z'=1$.

It has been demonstrated that the characteristic of secondary flow in oblique channels has great impact on the local, as well as global, heat transfer enhancement of the heat sink. Thus, it is important to examine the local coolant behaviour in the

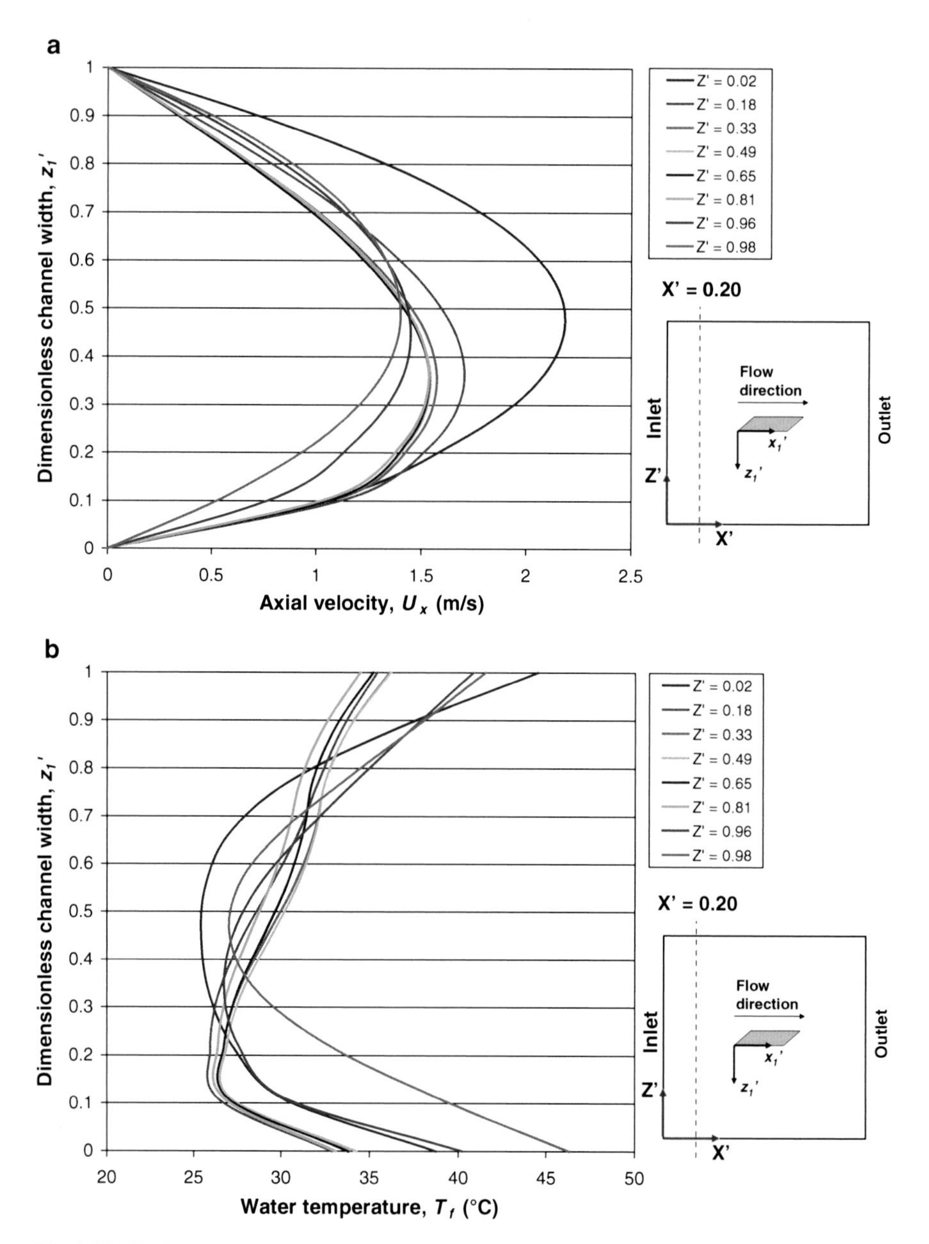

Fig. 2.25 Coolant (**a**) axial velocity and (**b**) temperature profiles in the main channels at $X'=0.20$, $Y'=0.5$, when $\mathrm{Re}=250$

oblique channels within the full domain heat sink. Figure 2.27a displays the coolant axial velocity profiles (mid-depth plane) in the oblique channels at different spanwise positions at the dimensionless axial distance of $X'=0.01$. At this upstream

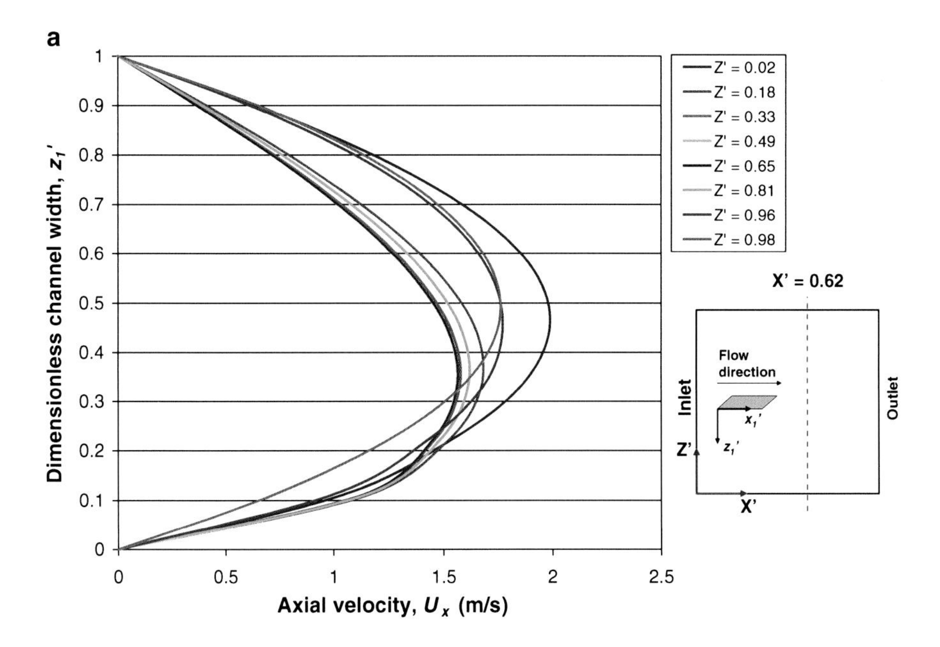

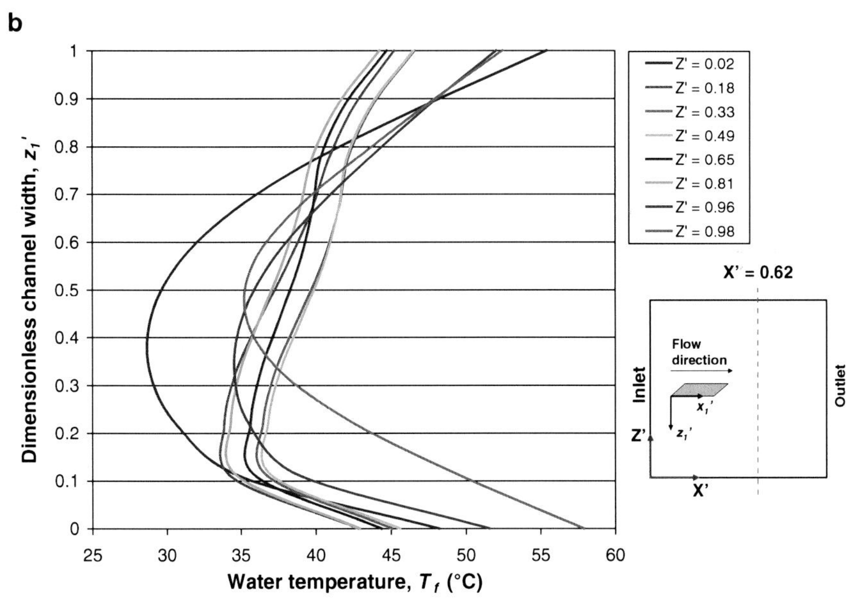

Fig. 2.26 Coolant (**a**) axial velocity and (**b**) temperature profiles in the main channels at $X'=0.62$, $Y'=0.5$, when $\mathrm{Re}=250$

position of the heat sink, there is a very limited amount of coolant mass diversion into the oblique channels. Axial velocity in most of the oblique channels is as slow as 0.1 m/s and displays a symmetrical profile at the centre of the channel.

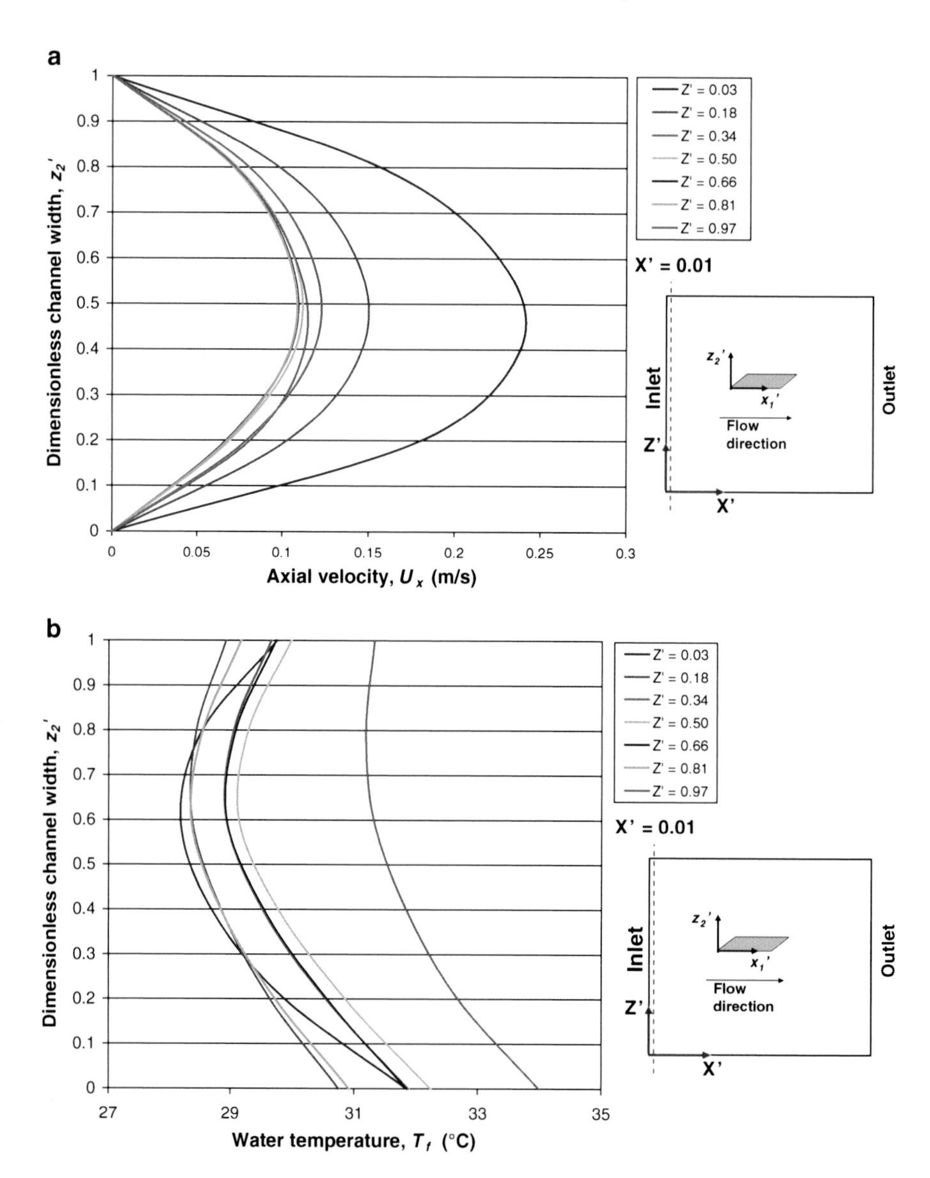

Fig. 2.27 Coolant (**a**) axial velocity and (**b**) temperature profiles in the oblique channels at $X'=0.01$, $Y'=0.5$ (*mid-depth plane*), when Re = 250

Despite the symmetry in velocity profiles in the oblique channels, water temperature profiles in Fig. 2.27b display highly asymmetric curves with the higher water temperatures occur for coolant at $z_2'<0.5$ and higher wall temperatures for fin surfaces $z_2'=0$ than those at $z_2'=1$. This is possible because these two fin surfaces at $z_2'=0$ and $z_2'=1$ belong to different oblique fins. Fin surface $z_2'=0$ is part of the downstream oblique fin that is located closer to the discrete heat source. Thus, more heat is expected to conduct into this fin and surface $z_2'=0$ in comparison with surface $z_2'=1$, which is one of the fin surfaces for the upstream fin.

As discussed earlier, significant changes occur to the velocity and temperature profiles in the main channels when coolant travels from $X'=0.01$ to $X'=0.20$. Thus, similar developments are expected for the profiles in the oblique channels. Except the two oblique channels that are located next to the heat sink edge ($Z'=0.03$ and $Z'=0.97$), the axial velocity in the oblique channels in Fig. 2.28a marks an appreciable rise. The magnitude of axial velocity increases with the distance from the

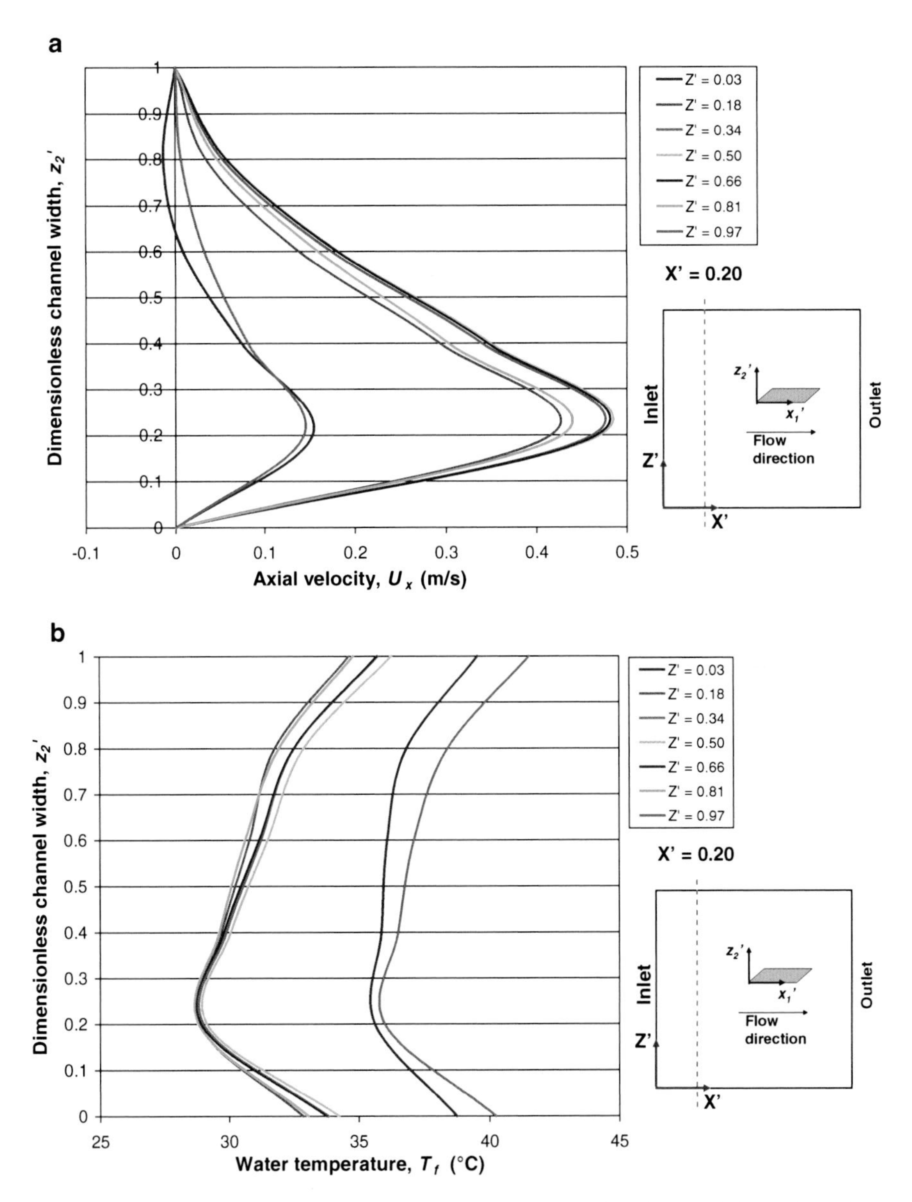

Fig. 2.28 Coolant (**a**) axial velocity and (**b**) temperature profiles in the oblique channels at $X'=0.20$, $Y'=0.5$, when Re = 250

heat sink edge, agreeing to the finding from Fig. 2.23. Therefore, coolant in the oblique channels between $0.18 < Z' < 0.81$ has significantly lower water and wall temperatures than those receiving less coolant as demonstrated in Fig. 2.28b. It is also noticed in Fig. 2.28 that these velocity profiles skew towards $z_2' = 0$ (fin surface **O1**), which subsequently produce a thinner boundary layer thickness for fin surface **O1** at $z_2' = 0$. Owing to the asymmetry of velocity profiles, temperature profile in the oblique channels displays the similar trend to that of the main channels, where the temperature profiles skew towards the fin surfaces with thinner hydrodynamics boundary layer. The thinner thermal boundary layer thickness on these fin surfaces results in superior heat transfer performance and also lower wall temperatures than the other fin surfaces. Detailed discussion on the heat transfer mechanism is present in the previous publication [22].

Comparable velocity and temperature profiles emerge further downstream in the oblique channels at $X' = 0.62$ as both the velocity and temperature profiles are noticed to stabilize as coolant flows past $X' = 0.20$. Again, the simulation findings show that flow migration affects the global secondary flow distribution within the heat sink but has little influence to the local velocity and temperature profiles. Apart from the channels that are located close to the heat sink edge, which is highly affected by flow migration and edge effect, channels within $0.18 < Z' < 0.81$ demonstrate consistent local velocity and temperature profiles for all axial distances examined (Figure 2.29).

Unlike the simulation findings in the previous analysis, where oblique fin microchannel heat sink achieves steady state condition at $Re = 250$, the simulation result shows signs of unsteadiness when Reynolds number is increased to $Re = 660$. Fluctuations are observed in velocity profiles and also in the temperature profiles in the heat sink. Figure 2.30a, b displays the local velocity and temperature profiles of liquid coolant at the mid-depth plane of a channel located at $X = 7.81$ mm, $Z = 8.28$ mm ($X' = 0.61$, $Y' = 0.5$, $Z' = 0.65$) when $Re = 250$. Based on the plots, it is observed that both the velocity and temperature profiles at different time steps overlap each other, indicating that the flow regime is steady. On the contrary, when examining the local velocity and temperature profiles of liquid coolant at the mid-depth plane of a channel at $X = 7.81$ mm, $Z = 4.19$ mm ($X' = 0.61$, $Y' = 0.5$, $Z' = 0.33$) when $Re = 660$ in Fig. 2.30c, d the fluctuation appear.

2.3.4 *Heat Transfer Performance*

In order to further explore flow migration phenomenon, local and global heat transfer performance was investigated based on the full domain simulation for oblique fin microchannel heat sink when $Re = 250$ and 660.

Figure 2.31 illustrates the temperature contour of heat sink (bottom wall) for the case $Re = 250$, where the wall temperature increases with the flow length due to the sensible heat gain by coolant and the variation of heat sink cooling performance. As this silicon-based microchannel heat sink has multiple discrete heat sources, the

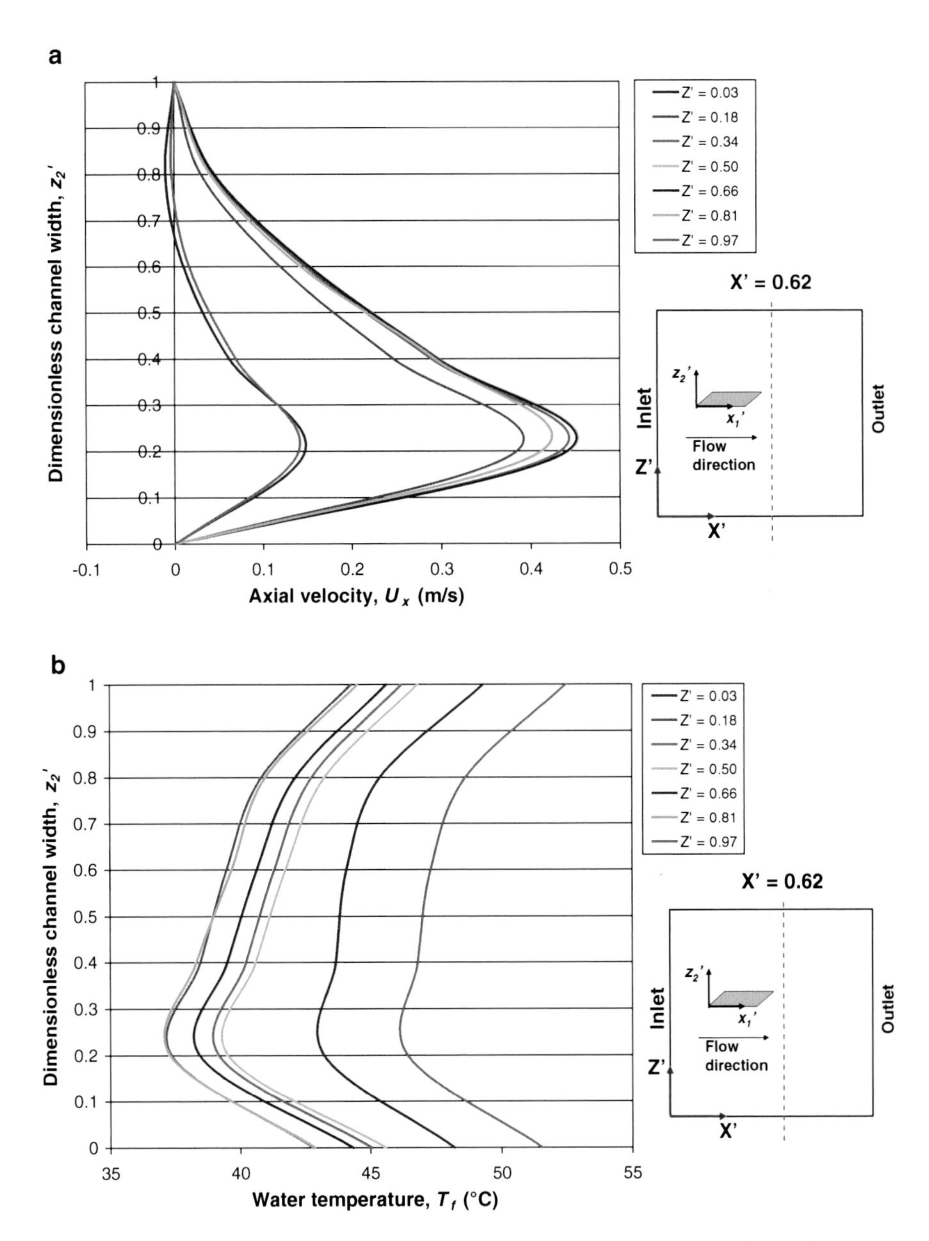

Fig. 2.29 Coolant (**a**) axial velocity and (**b**) temperature profiles in the oblique channels at $X'=0.62$, $Y'=0.5$, when Re = 250

highest temperature occurs within the heat source itself, and the wall temperature of non-heated area is obviously lower. It is also noticed that the maximum wall temperature of each individual heater is located slightly downstream from the centre of the heat source due to the combined effect of sensible heat gain by coolant, heat

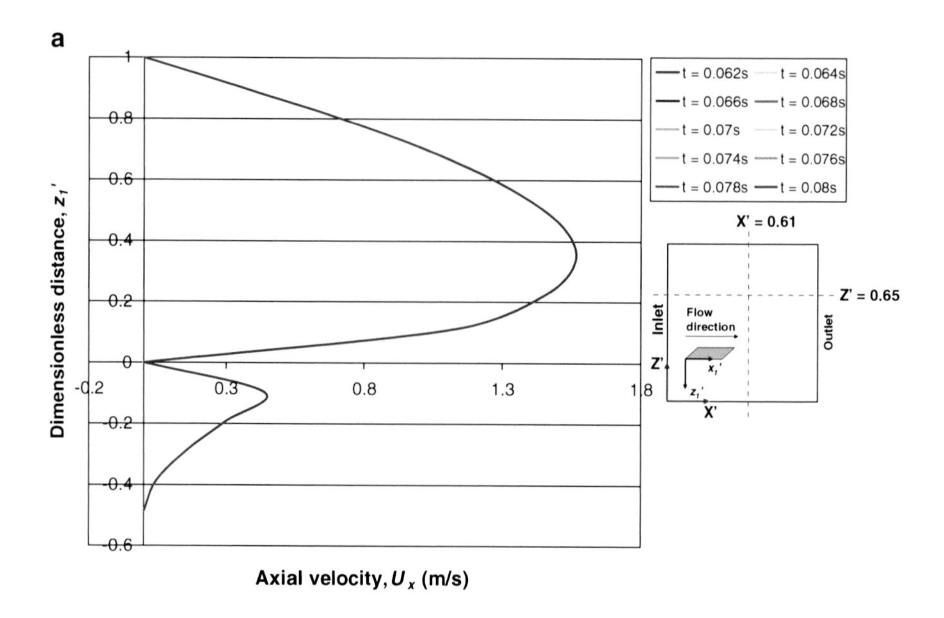

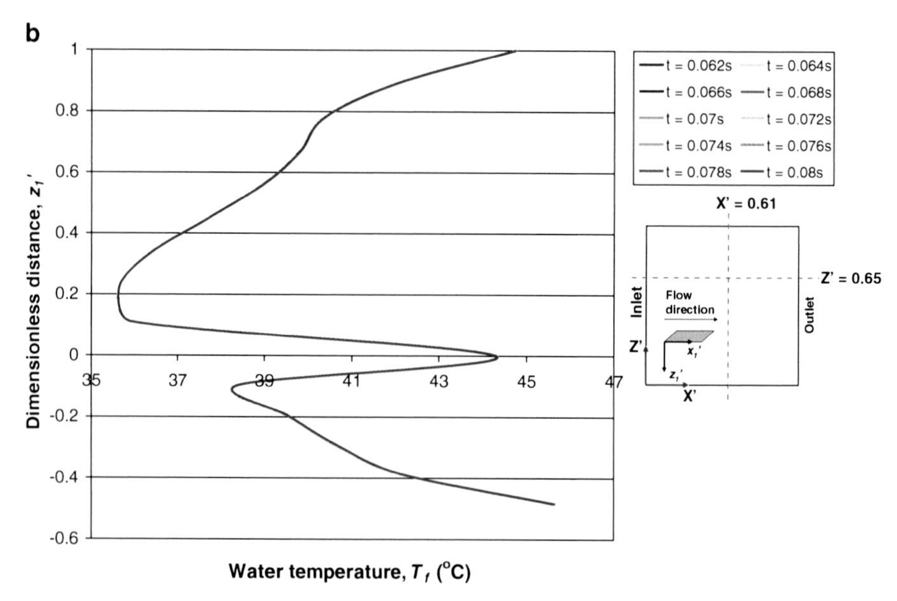

Fig. 2.30 Variation of (**a**) axial velocity profiles when Re = 250, (**b**) water temperature profiles when Re = 250, (**c**) axial velocity profiles when Re = 660, (**d**) water temperature profiles when Re = 660 in the mid-depth plane of channel within the oblique fin microchannel heat sink at different time steps

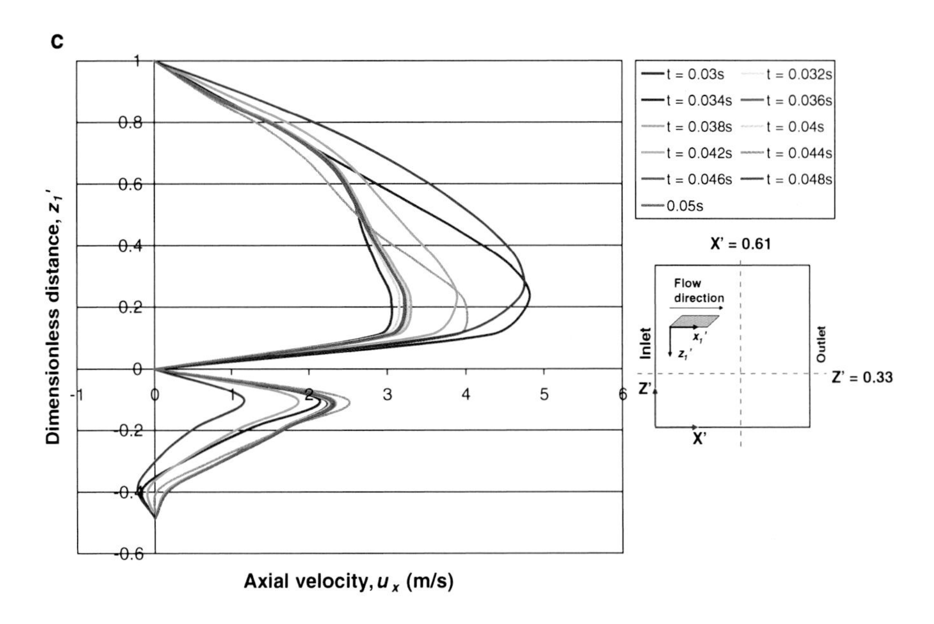

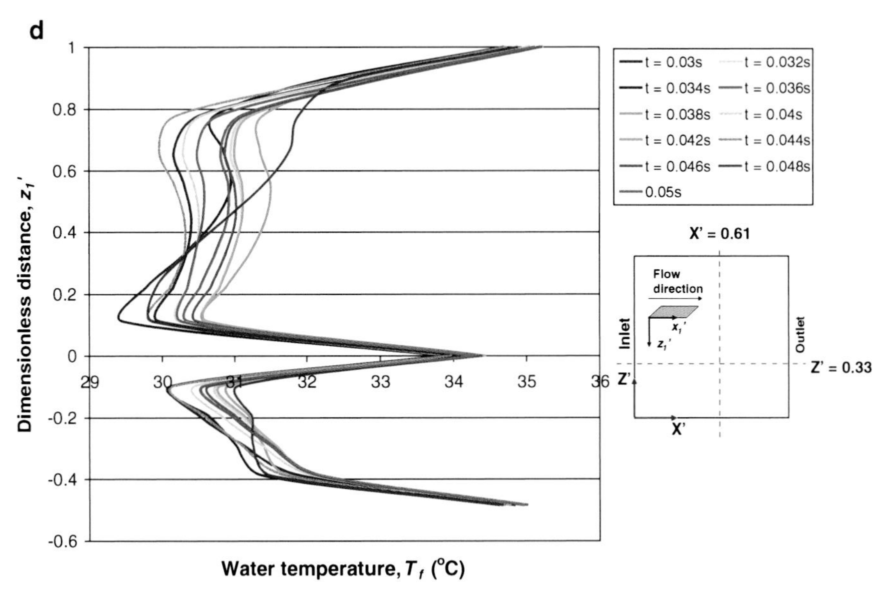

Fig. 2.30 (continued)

spreading and variation of heat transfer performance along the heat sink. Heaters on both edges of the heat sink ($Z'=0.1$ and $Z'=0.9$) also display higher wall temperatures than those at the centre of the heat sink. Similar trend is observed from Fig. 2.32, which plots the area-weighted average temperature of 25 heaters based on

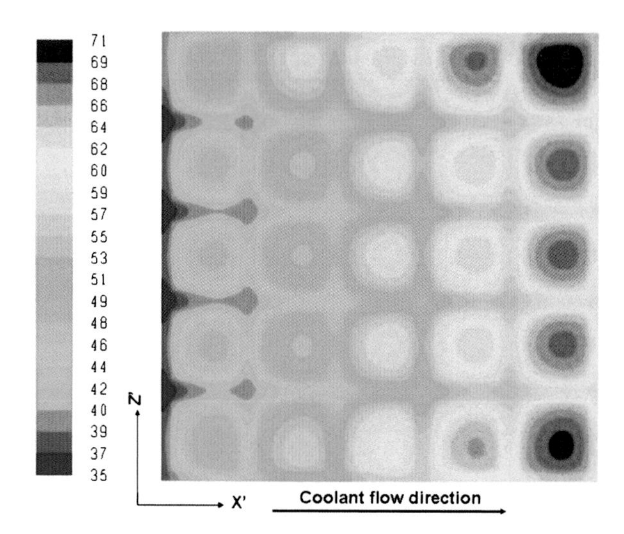

Fig. 2.31 Temperature contour (in °C) at the bottom wall of heat sink for Re = 250

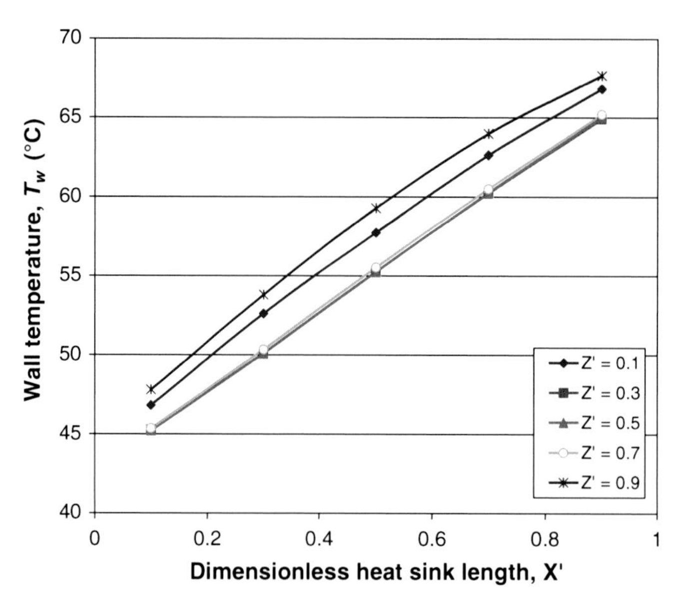

Fig. 2.32 Average heater wall temperature profiles in the oblique fin microchannel heat sink when Re = 250

the dimensionless heat sink width and axial distance. Wall temperatures are the highest for heaters at $Z' = 0.9$, followed by $Z' = 0.1$, while the lowest wall temperatures are recorded for heaters between $0.3 < Z' < 0.7$ and they almost overlap each other. Several contributing factors to such heater temperature profile are identified as non-uniform secondary flow distribution, flow migration, edge effect and heat spreading effect.

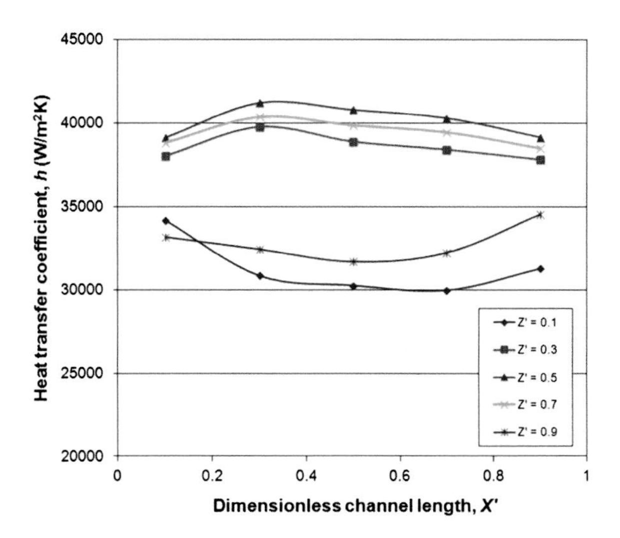

Fig. 2.33 Local heat transfer coefficient profiles of oblique fin microchannel heat sink when Re = 250

Figure 2.33 plots the local averaged heat transfer coefficients of the oblique fin microchannel heat sink in the streamwise direction for various spanwise positions when Re = 250. The centre zone of the heat sink, $Z'=0.5$, shows the highest heat transfer coefficients, followed by $Z'=0.7$ and $Z'=0.3$. The heat transfer coefficients at both edges of the heat sink, $Z'=0.1$ and $Z'=0.9$, are significantly lower than the centre zones. The magnitude of secondary flow rate has a prominent effect to the heat transfer performance of the oblique fin microchannel heat sink and subsequently its temperature contour or profile.

Due to the fact that coolant mass flow rate in the main channels varies in both spanwise and streamwise direction, sensible heat gain by coolant results in different local coolant temperature rise as shown in Fig. 2.34. The highest coolant flow rate results in the lowest coolant temperature in $Z'=0.1$, while the opposite is also true for $Z'=0.9$. This might explain that the heat sink at zone $Z'=0.9$ having higher wall temperature than zone $Z'=0.1$ although both having comparable secondary flow rates, heat transfer coefficient and heat transfer area. Besides, the total heat transfer area of these two zones is ~5 % lower than the others due to the edge effect (thicker fin width at both edges), which might have also contributed to the elevated edge temperatures.

For the case Re = 660, despite the unsteadiness in coolant flow, the bottom wall temperature of the heat sink is relatively consistent. Computed as an average temperature from 11 successive time steps, the maximum standard deviation of the heater temperatures is found at 0.15 °C. Figure 2.35 plots the area-weighted average heater temperature with the corresponding location on the heat sink. Compared with the average heater wall temperature profiles for low Reynolds number case shown in Fig. 2.36 where lowest coolant temperature occurs in $Z'=0.1$, heaters at both edges of the heat sink show higher wall temperature in the high Reynolds number case, which could be attributed to the severer flow migration. Subsequent heat transfer characteristics are also computed as an average of multiple time steps.

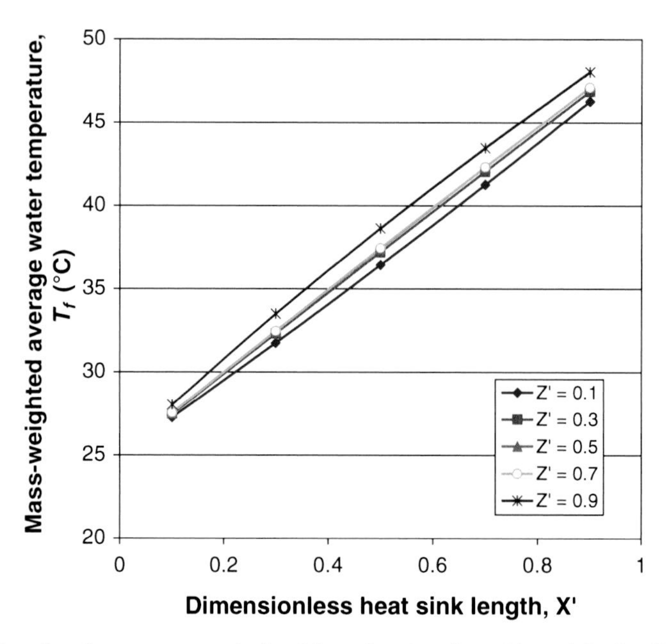

Fig. 2.34 Local coolant temperature in the oblique fin microchannel heat sink when Re = 250

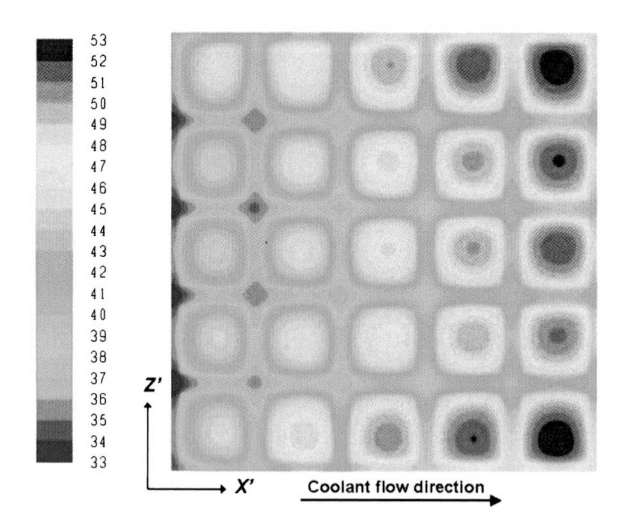

Fig. 2.35 Temperature contour (in °C) at the bottom wall of silicon-based oblique fin microchannel heat sink at Re = 660

Local heat transfer coefficient is then computed, as shown in Fig. 2.37. Employing the oblique fin microchannel heat sink helps to keep a very uniform heat transfer coefficient throughout the heat sink, in contrast to the conventional microchannel, where convective heat transfer performance deteriorates in the streamwise direction

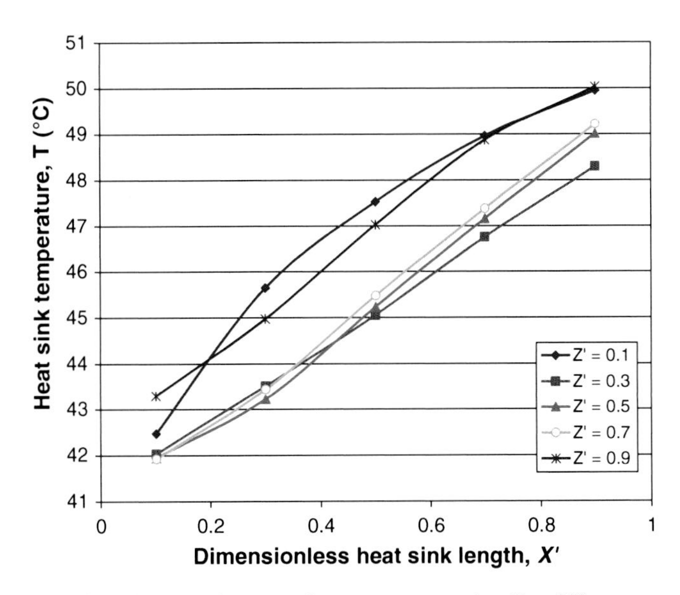

Fig. 2.36 Area-weighted average heater surface temperature when Re = 660

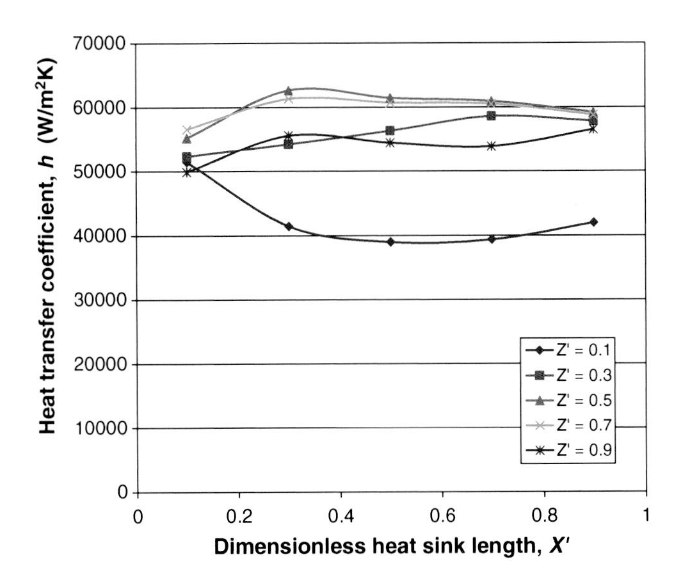

Fig. 2.37 Local heat transfer coefficients of oblique fin microchannel heat sink

due to boundary layer development [22]. Heat transfer coefficients for the oblique fin microchannel are kept within a narrow range for the high Reynolds number case, mostly between 50,000 and 60,000 W/m^{2}K, which can be comparable to two-phase boiling process. Similar to the low Reynolds number case, heat sink edges $Z' = 0.1$

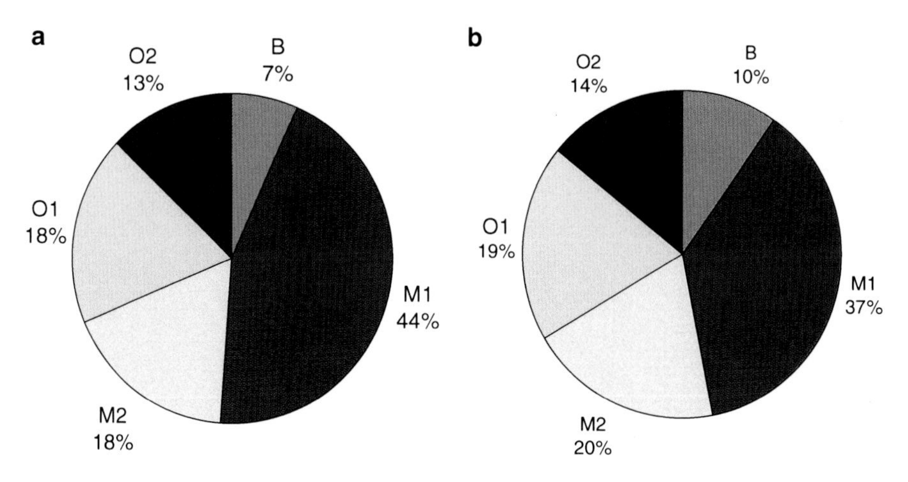

Fig. 2.38 Comparison of heat transfer rate between fin surfaces for (**a**) Re = 250; (**b**) Re = 660

still show lower heat transfer performance in comparison with the remaining part of the heat sink mainly due to significantly lower secondary flow generation. However, the local heat transfer performance in $Z' = 0.9$ is greatly enhanced due to the increased percentage of secondary flow rate.

The increment in Reynolds number elevates the secondary flow rate in oblique channels, affecting velocity profiles and subsequently heat transfer performance of each fin surface. Figure 2.38 compares the heat dissipation of each fin surface two cases considered in the study. Increment in secondary flow rate reduces the flow rate in the main channel. Thus, the effectiveness of fin surface M1 in dissipating heat is reduced. Higher secondary flow rate in the oblique channel improves the heat transfer performance in the oblique channel, as reflected by the increment in heat dissipation through surface O1 and O2. Subsequently, when secondary flow is injected to the adjacent main channel, higher fluid momentum thins the boundary layer at surface M2 further and improve the heat transfer of this fin surface.

2.4 Experimental Investigation: Average Performance

Simulation studies in the previous sections present a clear picture of the fluid flow characteristics and the heat transfer mechanism in the enhanced microchannel heat sink and the projected heat transfer augmentation. The focus of this section is shifted towards experimental investigation of the proposed heat transfer augmentation technique, using copper-based microchannel heat sinks. Both enhanced microchannel heat sinks of different materials and length scales demonstrate significant heat transfer improvement with negligible pressure drop up to a reasonably high Reynolds number (Re = 400–500). The experiment also results in good agreement with the simulation predictions, validating the simulation model [23].

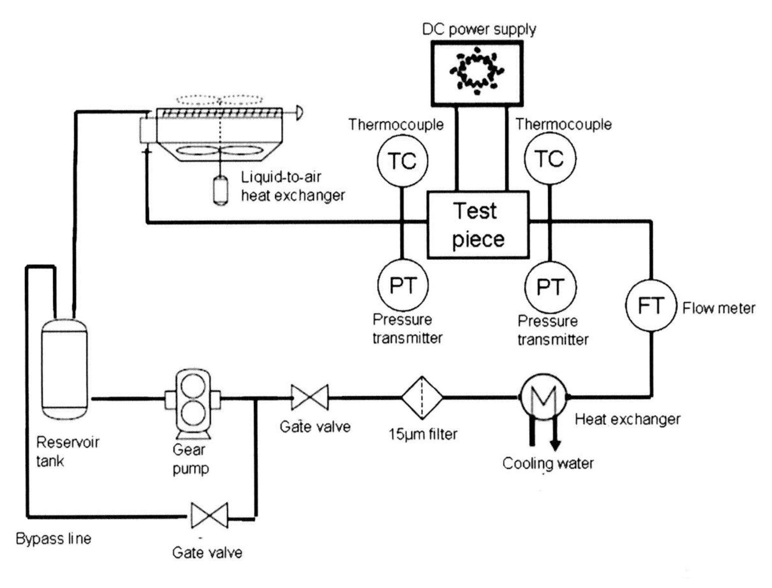

Fig. 2.39 Schematic for experimental flow loop [23]

2.4.1 Experimental Set-Up and Procedure

A schematic of the experimental flow loop is shown in Fig. 2.39. Deionized water from a reservoir tank is driven through the flow loop using a micro-annular gear pump. This pump forces the coolant through a 15 μm filter and a flow meter before entering the microchannel test section. A heat exchanger (connected to water bath) is used to regulate the water temperature before coolant enters the test section. Pressure transmitters are attached to the manifolds immediately before and after the test piece to measure the pressure drop across the test section. The water temperatures immediately before, $T_{f,\mathrm{in}}$, and after, $T_{f,\mathrm{out}}$, the microchannel heat sink in the manifolds are also measured with two T-type thermocouples (copper–constantan). Heated water that exits the test section is cooled by passing it through a liquid-to-air heat exchanger before it returns to the reservoir tank.

After the test section is assembled and properly sealed, the gear pump is switched on and the desired flow rate through the flow loop is controlled with the digital control on the gear pump. As the flow rate is stabilized, the power supply to the heaters is switched on, and steady state is usually reached in 30–45 min. The power input to the heaters in the test piece is controlled by a DC power supply unit. The voltage across the heaters is measured directly, while the high current through the heaters is calculated from Ohm's law based on the voltage measured across a shunt resistor that is connected in series with the heater. Input power to the heaters is calculated as the product of measured voltage and calculated current. Steady-state readings from the thermocouples, differential pressure transmitter and others are recorded by data logger and stored in a computer throughout the duration of the experiment. Each steady state value was calculated based on the average of 100 readings sampled at 0.1 Hz.

Two sets of copper-based microchannel heat sinks with nominal main channel width of 300 and 500 μm are included in the experimental studies. The measured geometrical dimensions for both sets of microchannel heat sinks are provided in Tables 2.3 and 2.4. In order to facilitate a fair performance comparison, both heat sinks are fabricated to a comparable channel aspect ratio, channel width, fin width and overall footprint. Measurement is performed with a 3-axis measurement microscope with 200x magnification at nine different points (3×3 grid) on the microchannel heat sink.

Details of the microchannel test section are shown in Fig. 2.40. The test section consists of a copper block, a Teflon manifold and a polycarbonate lid. The heat sink is machined from a square copper block of dimensions 25 mm×25 mm×70 mm. Microchannels are cut into the top surface by wire cutting, and the average surface roughness (Ra) measured using a white light interferometer surface profiler at the bottom wall of the channel is 1.81 μm. Holes are drilled into the bottom of the copper

Table 2.3 Dimensional details for microchannel heat sinks (300 μm nominal channel width)

Characteristic	Conventional microchannel	Enhanced microchannel
Material	Copper	
Footprint, width×length (mm)	25×25	
Number of fin row, n	39	
Main channel width, w_c (μm)	355.2	360.2
Fin width, w_w (μm)	245.8	245.2
Channel depth, H (μm)	1,153.1	1,174.0
Aspect ratio, α	3.25	3.26
Number of fin per row	–	20
Oblique channel width, w_{ob} (μm)	–	210.2
Fin pitch, p (μm)	–	1,164.5
Fin length, l (μm)	–	725.1
Oblique angel, θ (°)	–	26.6

Table 2.4 Dimensional details for microchannel heat sinks (500 μm nominal channel width)

Characteristic	Conventional microchannel	Enhanced microchannel
Material	Copper	
Footprint, width×length (mm)	25×25	
Number of fin row, n	22	
Main channel width, w_c (μm)	547.0	539.3
Fin width, w_w (μm)	458.5	465.2
Channel depth, H (μm)	1,482.2	1,487.1
Aspect ratio, α	2.71	2.76
Number of fin per row	–	12
Oblique channel width, w_{ob} (μm)	–	298.0
Fin pitch, p (μm)	–	1,995.2
Fin length, l (μm)	–	1,331.2
Oblique angel, θ (°)	–	26.4

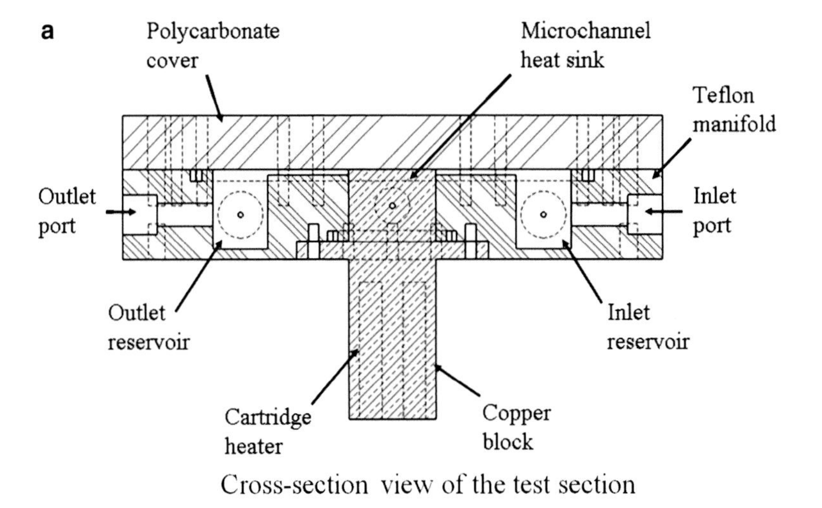

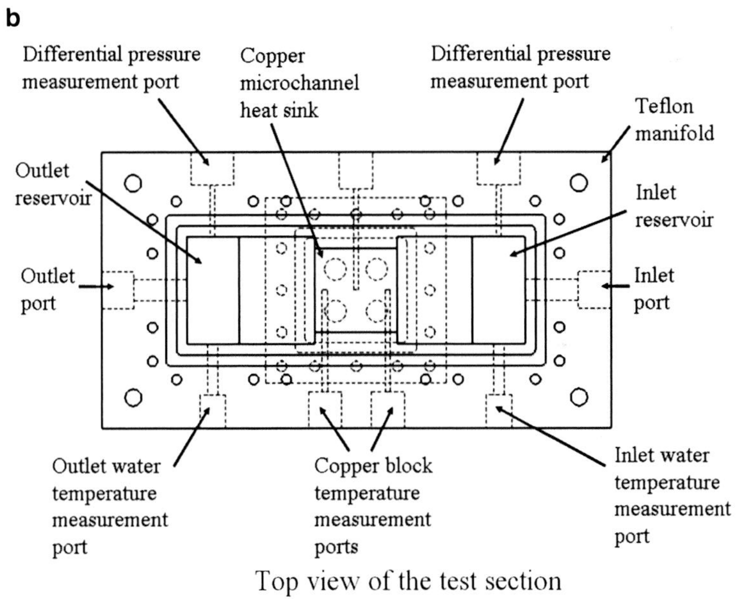

Fig. 2.40 Detailed drawings of the test section [23]. (**a**) Cross-section view of the test section. (**b**) Top view of the test section

block to accommodate four cartridge heaters that can provide a combined maximum power input of 1 kW. Three T-type (copper–constantan) thermocouple probes are positioned in the streamwise direction at a distance of 10 mm from the top surface of the copper block. The temperature readings from these thermocouple probes are extrapolated to provide an average bottom channel wall temperature of the heat sink.

This copper block is then mounted onto Teflon housing, where water tight seal (rubber O ring) is placed between the mating surfaces. The top surface of the copper block and Teflon manifold is then sealed with the polycarbonate cover. A thin layer of silicone rubber (250 μm) is laid between the microchannel heat sink and polycarbonate cover to ensure proper sealing.

The sensible heat gained by the coolant is determined from an energy balance:

$$q = \rho c_p Q \left(T_{f,\text{out}} - T_{f,\text{in}} \right) \tag{2.20}$$

It is found that more than 95 % of the heat input is transferred to the coolant across all the experimental runs, where the heat loss is deemed minimum or negligible. The heat flux into the test vehicles is calculated as the sensible heat gain into coolant over the area of copper block. An average heat flux at 100 W/cm^2 is maintained throughout the experiments for the copper-based microchannel heat sinks with 300 μm nominal channel width, while lower heat flux level at 65 W/cm^2 is supplied into heat sink with 500 μm nominal channel width in order to have the flow condition maintained in single phase flow throughout the experiment.

The average heat transfer coefficient can be then determined using the correlation:

$$h = \frac{q}{A_{\text{tot}} \left(T_w - T_{f,\text{ave}} \right)} \tag{2.21}$$

where A_{tot} is total area of convective heat transfer surfaces. For conventional microchannel heat sink, total heat transfer area is evaluated as

$$A_{\text{tot}} = (N+1) L \left(w_{\text{ch}} + 2\eta H \right) \tag{2.22}$$

On the other hand, total area of heat transfer for oblique finned microchannel heat sink is calculated as

$$A_{\text{tot}} = A_b + \eta A_{\text{fin}} \tag{2.23}$$

where A_b is the unfinned surface area at the bottom of the channels and A_{fin} is fin area.

A_b = (Heat sink base area) − (Base area occupied by oblique fins)

$$A_b = WL - N w_w l \tag{2.24}$$

$$A_{\text{fin}} = NHP \tag{2.25}$$

For microchannel configuration, fin efficiency is used to account for the drop in temperature along the fin. An adiabatic fin tip condition is assumed due to the insulative material of the manifold cover, and the corresponding fin efficiency is given as

$$\eta = \frac{\tan h(mH)}{mH} \tag{2.26}$$

where $m = \sqrt{\frac{hP}{k_{\text{HS}} A_c}}$

As direct wall temperature measurement at the bottom of channel is not available, extrapolation from the temperatures measured by the thermocouples in the copper heat sink is performed by assuming 1-D heat conduction:

$$T_w = T_{\mathrm{HS}} - \frac{sq''}{k_{\mathrm{HS}}} \tag{2.27}$$

HS denotes heat sink. On the other hand, $T_{f,\mathrm{ave}}$ is the average fluid temperature. The corresponding Nusselt number is then calculated as $\mathrm{Nu} = \frac{hD_h}{k_f}$, where D_h is the hydraulic diameter of the channel and k_f is the thermal conductivity of water.

Total thermal resistance of the heat sink is defined as

$$R_{\mathrm{tot}} = \frac{T_{max} - T_{f,\mathrm{in}}}{q} \tag{2.28}$$

where $T_{\max}$ is the maximum measured temperature of the heat sink, T_f is the inlet coolant temperature and q is the heat supplied into the heat sink.

As for material properties, copper is assumed to have constant thermal conductivity, k_{cu} = 387.6 W/mK. The density, specific heat capacity, thermal conductivity and dynamic viscosity of water are evaluated at the mean fluid temperature (average of the fluid inlet and outlet temperatures) based on the formulas stated in the previous section.

As the pressure transmitters are located at the manifolds, the pressure drop measurement represents the combined losses due to the frictional loss in microchannels and minor losses due to abrupt contraction and expansion at the inlet and outlet [24], which can be written as follows:

$$\Delta P = \Delta P_{c1} + \Delta P_{c2} + \Delta P_{ch} + \Delta P_{e2} + \Delta P_{e1} \tag{2.29}$$

The pressure drop across microchannel can then be calculated as

$$\Delta P_{ch} = \Delta P - \left(\Delta P_{c1} + \Delta P_{c2} + \Delta P_{e2} + \Delta P_{e1}\right)$$

where ΔP_{c1} and ΔP_{c2} are the contraction pressure losses from the deep plenum to the shallow plenum and from the shallow plenum to the microchannel. These minor losses can be expressed as [25]

$$\Delta P_{c1} = \frac{1}{2}\rho_f\left(u_{p2,\mathrm{in}}^2 - u_{p1,\mathrm{in}}^2\right) + \frac{K_{c1}}{2}\rho_f u_{p2,\mathrm{in}}^2 \tag{2.30}$$

$$\Delta P_{c2} = \frac{1}{2}\rho_f\left(u_{\mathrm{in}}^2 - u_{p2,\mathrm{in}}^2\right) + \frac{K_{c2}}{2}\rho_f u_{\mathrm{in}}^2 \tag{2.31}$$

where $p1$ and $p2$ denote the deep plenum and shallow plenum, respectively and K_{c1} and K_{c2} are the loss coefficients for the abrupt contractions. On the other hand, ΔP_{e2} and ΔP_{e1} express the pressure losses from the microchannel to the shallow plenum and from the shallow plenum to the deep plenum, which can be written as follows:

$$\Delta P_{e2} = \frac{1}{2}\rho_f\left(u_{p2,\text{out}}^2 - u_{\text{out}}^2\right) + \frac{K_{e2}}{2}\rho_f u_{\text{out}}^2 \quad (2.32)$$

$$\Delta P_{e1} = \frac{1}{2}\rho_f\left(u_{p1,\text{out}}^2 - u_{p2,\text{out}}^2\right) + \frac{K_{e1}}{2}\rho_f u_{p2,\text{out}}^2 \quad (2.33)$$

where K_{e1} and K_{e2} represent the loss coefficients due to the abrupt expansion. For the present heat sink test section geometry, the value of K_{c1}, K_{c2}, K_{e1} and K_{e2} is close to unity.

The associated friction factor is given by

$$f = \frac{\Delta P_{ch} D_h}{2\rho u_m^2 L} \quad (2.34)$$

The experimental investigation on heat sinks with 500 μm channel width is conducted over the flow rates ranging from 375 to 950 mL/min, which correspond to Reynolds numbers of 325–780. Similar coolant flow rate was applied for heat sinks with 300 μm channel width resulting Reynolds numbers from 260 to 640. Using $L/\text{Re}D_h > 0.05$ and $L/\text{Re}D_h\text{Pr} > 0.05$ as the criteria for fully developed flow, all the data points fall into the thermally developing regime.

2.4.2 Microchannel Heat Sinks with 500 μm Nominal Channel Width

Figure 2.41 plots the average Nusselt number to Reynolds number for both conventional and enhanced microchannel heat sinks. Generally, the average Nusselt number for both configurations increases with Reynolds number as the thermal boundary layer thickness decreases with the increased fluid velocity. However, the heat transfer for the enhanced microchannel with oblique fins is significantly higher compared with conventional microchannel heat sink. At the lowest Reynolds number of 325, the average Nusselt number is increased by 57 %, from 9.1 to 14.3. This appreciable enhancement in heat transfer is due to the combined effects of thermal boundary layer redevelopment at the leading edge of each oblique fin and the secondary flows generated by flow diversion through the oblique channels. The current experimental results are in accordance with the findings from the simulation presented in the previous chapters.

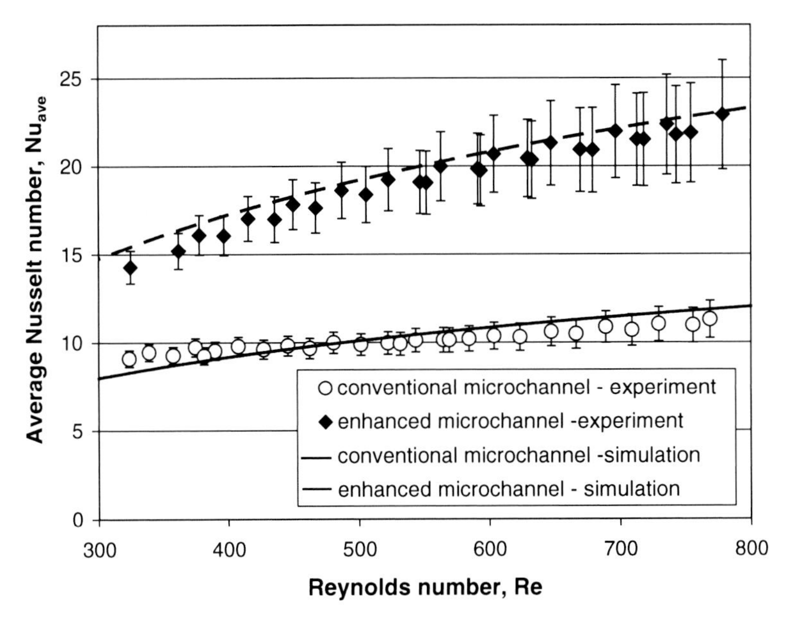

Fig. 2.41 Comparison of average Nusselt number for microchannel heat sinks (500 μm nominal channel width) [23]

The percentage of heat transfer augmentation is observed to increase with Reynolds number. A heat transfer enhancement of 103 % is achieved at the highest Reynolds number of 780, when the average Nusselt number increases from 11.3 to 22.9. As flow rate is increased, the water coolant would flow with a higher velocity through the channels. Coupled with higher percentage of flow diversion from the main channel into the oblique channel, the resultant secondary flows would carry the higher momentum and further disrupt the thermal boundary layer development and better mix the fluid leading to higher heat transfer enhancement at larger Reynolds number. Besides experimental data, simulation results of both conventional and enhanced microchannel are plotted in Fig. 2.41 for comparison purpose. These simulations adopt the simplified model proposed in Sect. 2.2. Only a pair of fin-channel is simulated for both enhanced microchannel and conventional microchannel configurations, where the spanwise repeating channels/fins are represented by either periodic or symmetry boundary condition. In fact, the simulation predictions are in good agreement with the experimental results, for both conventional and enhanced microchannel. It is noticed that all the experimental data falls within ±10 % of the predicted value, thus demonstrating the ability of the numerical models to predict the heat transfer performance especially for the enhanced microchannel heat sink, which adopted the simplified periodic boundary condition.

For predicting the heat transfer performance for louvred fin heat exchanger under constant heat flux condition, Aoki et al. [6] proposed an empirical correlation as below.

$$Nu_{ave} = 0.87\,Re_L^{0.5}\,Pr^{1/3} \tag{2.35}$$

This correlation yields an average Nusselt number that is merely 4 % lower than the Pohlhausen solution for laminar flow over a flat plate with constant heat flux. When this correlation is used on oblique finned microchannel heat sink, it over-predicts the heat transfer coefficient by averagely 44 %. Drastically different flow field between louvred fin heat exchanger and oblique finned microchannel heat sink is believed to be the source of discrepancy. As discussed in the previous sections, heat transfer efficiency through each oblique fin surface is significantly different from each other due to the asymmetrical coolant velocity and temperature profiles. In addition, the applicability of Pohlhausen solution in louvred fin heat exchanger shows that boundary layers are totally renewed at each louvre. The large distance between louvres promotes the thermal wake dissipation in the louvred fin heat exchanger [26], while oblique fins are closely packed and the wake dissipation is incomplete over the short wake length.

The experimental pressure drop data across microchannel heat sinks is presented in Fig. 2.42, alongside the simulation results, again showing very good agreement. It is interesting to note that the pressure drop across the enhanced microchannel is comparable to conventional microchannel for a Reynolds number below 400.

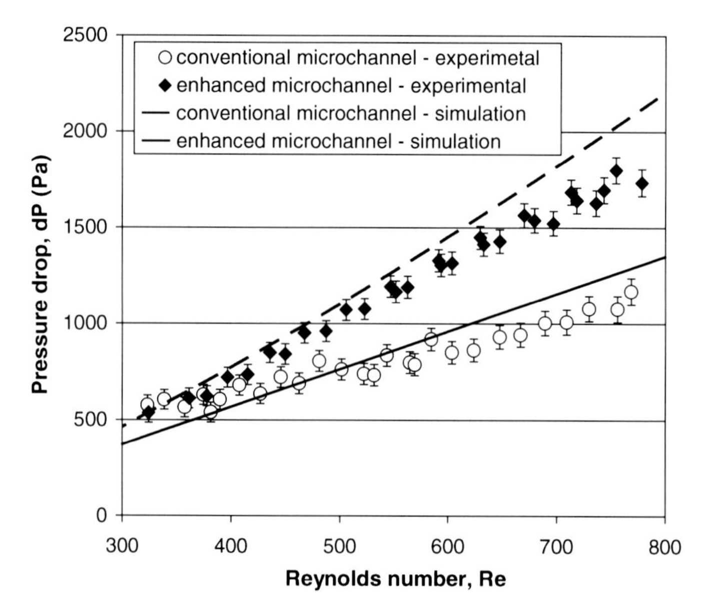

Fig. 2.42 Comparison of pressure drop across microchannel heat sinks (500 μm nominal channel width) [23]

This suggests that the enhanced heat transfer can be achieved without incurring excessive pressure drop penalty. This phenomenon is possible due to the competing effects of the increased pressure drops (due to thinning of boundary layers and boundary layer separation) and pressure recoveries (due to flow bifurcations) [14]. However, as the Reynolds number increases beyond 400, a greater amount of the flow will be diverted into the oblique channels. This creates secondary flows with stronger momentum, which further augment the heat transfer but incur some pressure drop penalty. The pressure drop penalty, however, is small compared to the heat transfer enhancement factor.

Heat transfer enhancement factor (E_{Nu}) and pressure drop penalty (E_f) are defined as the average Nusselt number and friction factor of the present enhanced microchannels divided by that of conventional straight channels, respectively [27]. Similar performance factor was adopted by Sui et al. [28]. From Fig. 2.43, the advantage of oblique finned microchannel is clearly revealed where the heat transfer enhancement factor outweighs the pressure drop penalty. When pressure drop penalty is close to 1 (no additional pressure drop incurred), the heat transfer performance (in terms of Nusselt number) of the oblique finned microchannels is 50–60 % more than that of conventional microchannels. As the Reynolds number increases, the heat transfer enhancement factor may exceed 2 (100 % improvement) while incurring a smaller pressure drop penalty of 40–60 %. However, it should be noted that the magnitude of pressure drop penalty is less than 1 kPa, which is relatively low and should be manageable with the same pump.

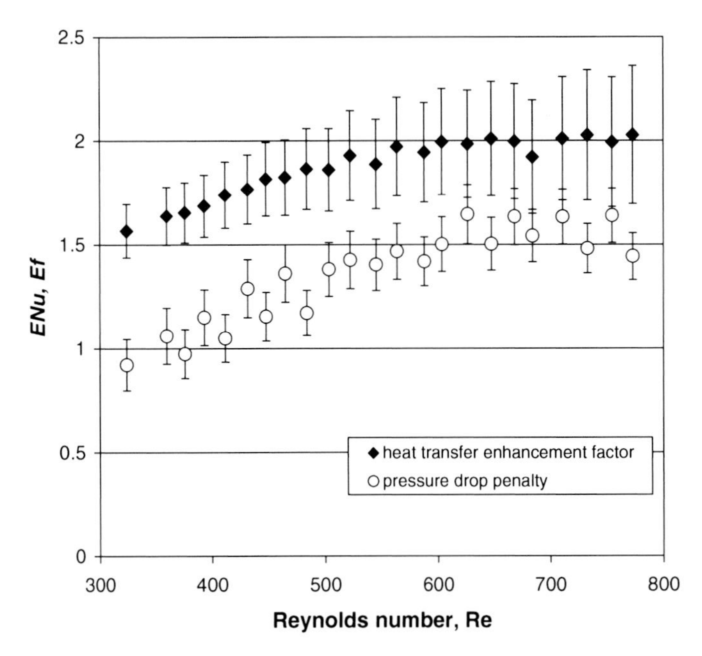

Fig. 2.43 Comparison of heat transfer enhancement factor and pressure drop penalty (500 μm nominal channel width) [23]

2.4.3 Microchannel Heat Sink with 300 μm Nominal Channel Width

Experiment investigation into heat sinks with smaller nominal channel and fin width at 300 μm also indicates significant heat transfer augmentation. This shows the applicability of the proposed heat transfer enhancement technique on heat sinks with different fin-channel size, both larger and smaller fin-channel size. A 77 % improvement is realized at Reynolds number as low as 260, where the average Nusselt number is elevated from 6.6 of the conventional microchannel to 11.7. The heat transfer performance of the enhanced microchannel by 111 % over the conventional configuration when the Reynolds number is raised to 650. Figures 2.44 and 2.45 show the comparison for Nusselt number and pressure drop between the conventional microchannel and enhanced microchannel, respectively.

Unfortunately, the results are confounded by the occurrence of bent/burr at the edge of oblique fins that points towards the main channel as displayed in Fig. 2.46. According to Joshi and Webb [29] in their study on offset strip-fin heat exchanger, the "burred edge" of fins could lead to 10–20 % increment in friction factor and also adversely affect the heat transfer. For future research, better fabrication technique has to be identified for microchannel heat sink with fin width less than 500 μm to reduce burr at the fin edge that could affect the performance of heat sink.

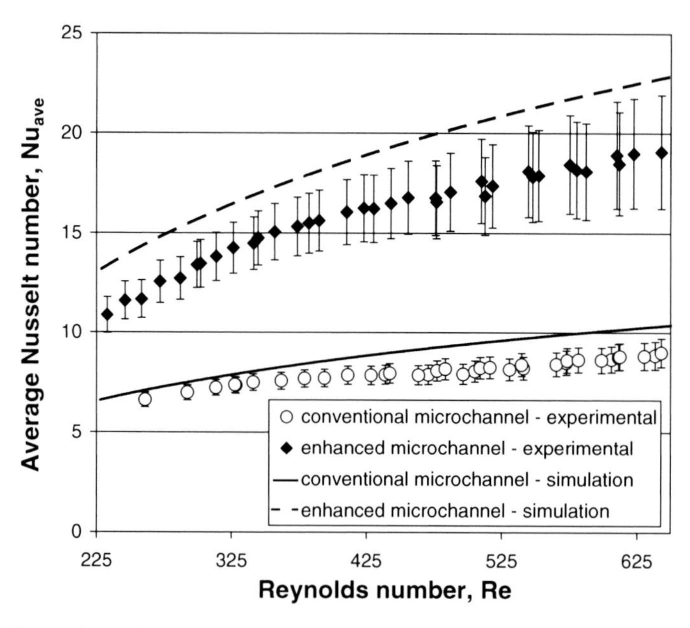

Fig. 2.44 Comparison of average Nusselt number for microchannel heat sinks (300 μm nominal channel width) [23]

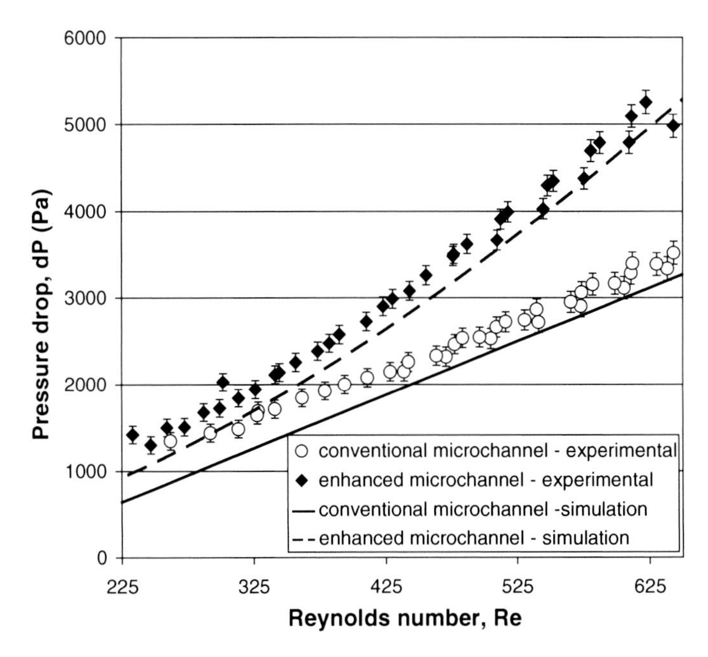

Fig. 2.45 Comparison of pressure drop across microchannel heat sinks (300 μm nominal channel width) [23]

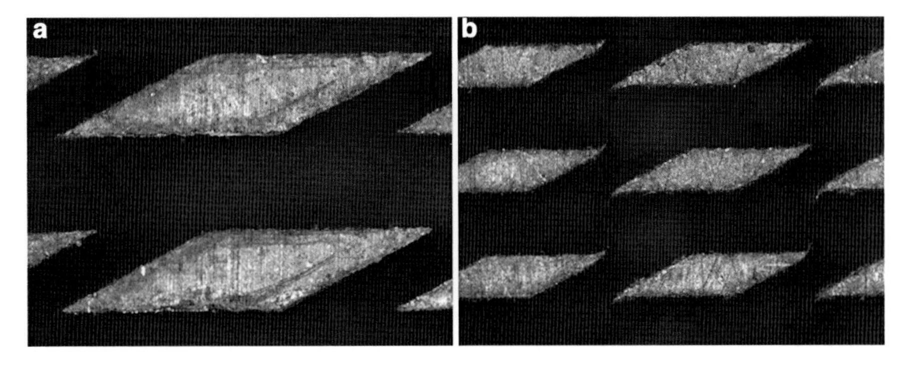

Fig. 2.46 Microscopic image of (**a**) enhanced microchannel heat sink with 500 μm nominal channel width (**b**) enhanced microchannel heat sink with 300 μm nominal channel width [23]

2.5 Experimental Investigation: Local Performance

The experimental investigation of copper-based oblique fin heat sink shows promising results. However in a recent review, Kandlikar et al. [30] observed that the heat transfer coefficients of these oblique finned copper microchannels were relatively low due to large hydraulic diameters and it will be interesting to see performance at smaller level [21].

2.5.1 Experimental Set-Up and Procedure

A similar set-up as used for the copper microchannel is used for the silicon-based microchannel heat sink. Here, silicon-based microchannel heat sinks are created on flip chip packages for experimental investigation. Channels are cut on the backside of the silicon thermal test chips using wafer dicing saw blades of different width. Measurement is performed with a 3-axis measurement microscope with 200x magnification at nine different points (3×3 grid) on the microchannel heat sink. The average surface roughness (Ra) measured at the bottom wall of the channel is 178 nm. Figure 2.47a displays the test section of microchannel heat sinks in this experiment. It consists of a polycarbonate manifold that is mounted onto a flip chip package, where the microchannels are laid. Each thermal test die is 0.1″×0.1″ (2.54 mm×2.54 mm) in size and when diced in 5×5 grid array, this results in an overall footprint of 0.5″×0.5″ (12.7 mm×12.7 mm) with 0.65 mm thickness for the current test vehicles. Each thermal test die has a heater (doped silicon well) at the bottom wall of the chip at 2 mm×2 mm and a series of thermal sensors (five diodes connected in series), which enable independent heater control and local temperature measurement, providing greater insight into local heat transfer behaviour and the overall temperature mapping. The resistive heating of the silicon thermal test dies is accomplished by driving the current through the doped silicon well, where the input power is controlled by a DC power supply unit. Figure 2.47b, on the other hand, shows the test piece and the enlarged view of the 5×5 arrays of thermal test dies. A coordinate numbering scheme (*X*, *Z*) for thermal test dies identification is also indicated in the figure.

Temperatures on the thermal test chip are indicated as voltage drop across the thermal diode sensors. Prior to the experiment, the voltage-temperature response of these thermal diode sensors are established through calibration. The calibration is performed in a convection oven from 30 to 90 °C, in steps of 10 °C. Temperatures and voltage drop are then recorded when both temperatures of oven and thermal sensors reach steady state, typically in an hour time.

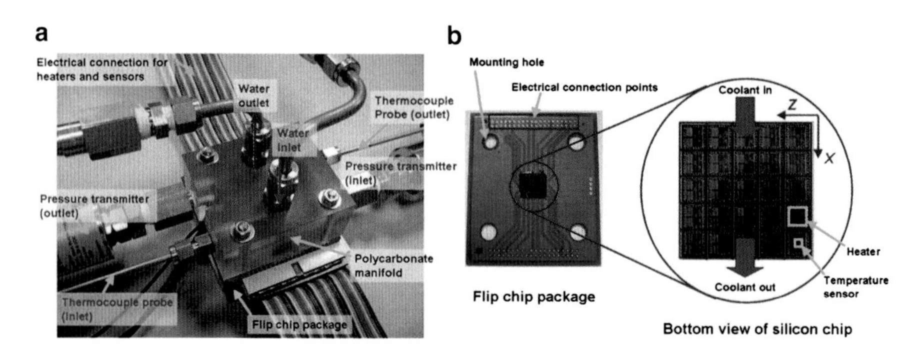

Fig. 2.47 (**a**) Microchannel test section; (**b**) test piece with 5×5 array of thermal test dies

2.5.2 Silicon-Based Microchannel Heat Sink with 100 μm Nominal Channel Width

The experiments are conducted over the flow rates ranging from 100 to 500 mL/min, which correspond to Reynolds numbers of 180–680. By employing silicon thermal test chip in these experiments, the local behaviour (local temperature and local heat transfer coefficient profiles) of the enhanced heat sink can be investigated. In this case, the performance of the enhanced microchannel heat sink with 100 μm nominal width under the experimental conditions of 160 mL/min (Re = 252) total coolant flow rate and total 273 W heater power of is presented.

The enhanced microchannel shows significant reduction in total thermal resistance compared to the conventional microchannel. The total thermal resistance comprises of conductive, spreading, convective and caloric thermal resistances. Generally, the conductive thermal resistance remains constant while spreading, convective and caloric thermal resistances reduce with the increasing Reynolds number, resulting in lower total thermal resistance. The improved heat transfer performance of the enhanced microchannel is demonstrated in Fig. 2.48, where the total thermal resistance of the enhanced microchannel heat sink is consistently lower than the conventional configuration in the range of Reynolds numbers studied. At low Reynolds number ~180, the total thermal resistance of the enhanced microchannel is 3 % (R_{tot} = 0.234 °C/W) lower than that of the conventional microchannel (R_{tot} = 0.242 °C/W). As Reynolds number rises, the effectiveness of enhanced microchannel increases, and the percentage of reduction in total thermal resistance quickly increases. The maximum total thermal resistance reduction achieved at Reynolds number ~690 is as much as 25 % ($R_{tot,EM}$ = 0.089 °C/W versus $R_{tot,CM}$ = 0.119 °C/W). In the context of current experiment, the highest percentage of

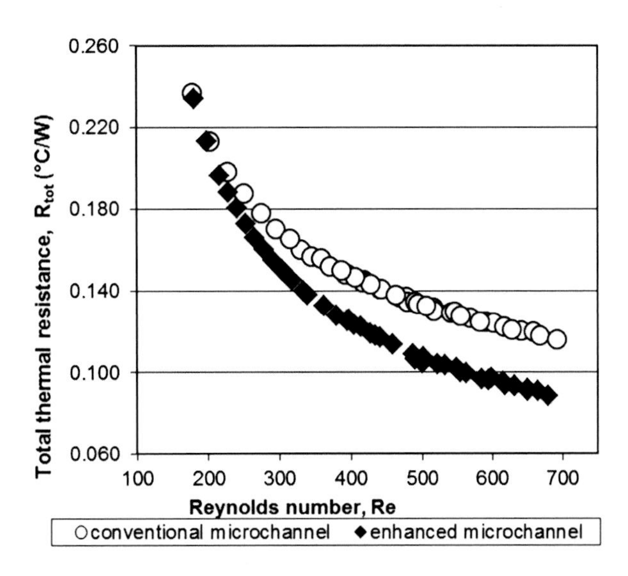

Fig. 2.48 Comparison of total thermal resistance between the microchannel heat sinks

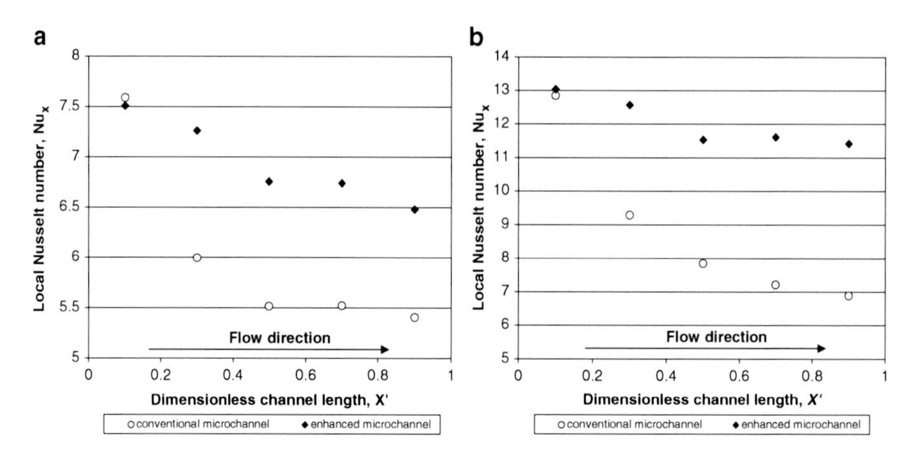

Fig. 2.49 Comparison of local heat transfer performance of microchannel heat sinks for (**a**) Re ~250; (**b**) Re ~680

total thermal resistance reduction translates to an 8.1 °C lower in maximum chip temperature for the enhanced microchannel heat sink ($T_{\max,EM}=50.5$ °C versus $T_{\max,CM}=58.6$ °C).

As discussed, the introduction of oblique fins and channels leads to a much uniform heat removal capability across the microchannel heat sink. Figure 2.49a, b illustrates the comparison of local heat transfer performance of microchannel heat sinks for two of the experimental runs at Re ~250 and Re ~680. The markers denote the location of thermal diode sensors on the silicon test chip in the streamwise direction, where the temperature measurement is made. It is noticed that both conventional and enhanced microchannel heat sinks display a relatively close Nusselt number at the upstream of the heat sink, $X'=0.1$ when the boundary layers are thin. The distinguishing point for these two heat sinks is the development of the flow regime and boundary layers as coolant travels downstream. As boundary layers continue to thicken with the flow distance, heat transfer performance of the conventional microchannel deteriorates rapidly, as showed in the figures for both Reynolds number cases. A maximum 48 % drop in Nusselt number is observed between $X'=0.1$ and $X'=0.9$, resulting in a highly non-uniform heat transfer performance within the heat sink. Instead of declining, the enhanced microchannel keeps the Nusselt number at a much elevated level and consistent across the enhanced microchannel heat sink. The profile of the streamwise local Nusselt number is very uniform, confined to a narrow range of 7.0–8.2 and 12.4–14.4 for both Reynolds number cases, respectively, from the inlet of the microchannel to its outlet. This phenomenon is highly due to the frequent re-initialization of thermal boundary layers and generation of secondary flows. This combination ensures that the flow is in a constant state of development thus having a sustainable performance close to that of the flow upstream.

The average heat transfer performance of the microchannel heat sinks is plotted in Fig. 2.50. Generally, the average Nusselt number, Nu_{ave}, increases with Reynolds

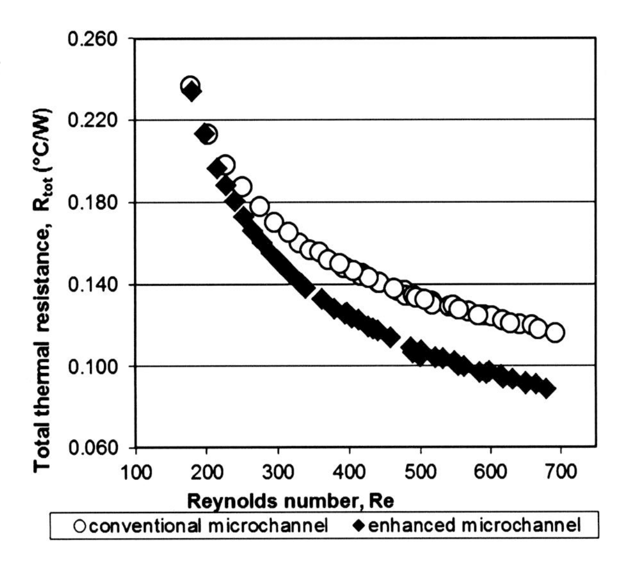

Fig. 2.50 Comparison of heat transfer performance for microchannel heat sinks

number as the thermal boundary layer thickness decreases with the increased fluid velocity. However, the heat transfer for the enhanced microchannel with oblique fins is highly augmented in comparison with conventional microchannel heat sink. At Reynolds number ~180, the average Nusselt number of the enhanced microchannel is 15 % higher than that of conventional microchannel. As Reynolds number rises, the Nusselt number increases by almost 47 %, from 9.0 to 13.2. The appreciable enhancement in heat transfer is due to the combined effects of thermal boundary layer redevelopment at the leading edge of each oblique fin and the secondary flows generated by flow diversion through the oblique channels. As the flow rate induced through the heat sink is increased, more fluid is diverted from the main channel into the oblique channel. This secondary flow thus carries higher momentum and further disrupts the boundary layers development and augments the heat transfer.

Nevertheless, the most interesting feature of the enhanced microchannel heat sink is that the significant heat transfer augmentation is achieved with small pressure drop penalty. The enhanced microchannel heat sink, which employs secondary flow to enhance its heat transfer, still managed a comparable pressure drop with the conventional microchannel heat sink for Reynolds number lower than 500, as displayed in Fig. 2.51. This distinguishes the proposed scheme from the conventional heat transfer enhancement scheme, where trade-off in terms of pressure drop penalty is inevitable. As the Reynolds number increases, a higher percentage of coolant will be diverted into oblique channels. This creates a secondary flow with stronger momentum, which further augments the heat transfer but incurs additional pressure drop penalty. Therefore, the pressure drop for enhanced microchannel starts to deviate and increases more than the conventional configuration. However, the magnitude of increment of pressure drop is considered manageable for the same micropump.

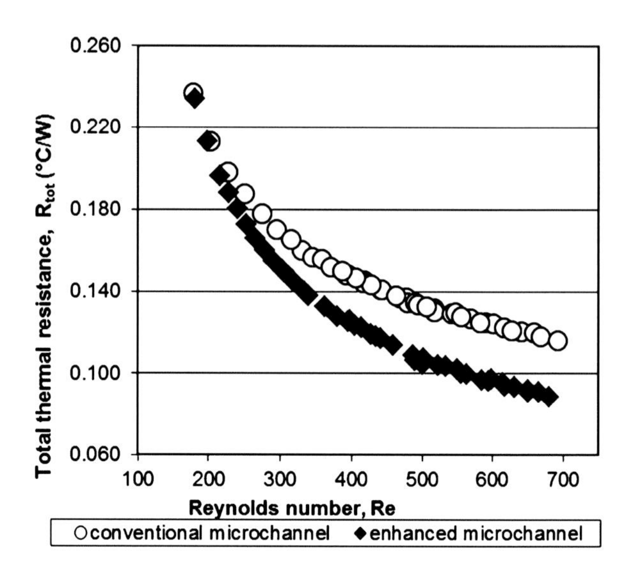

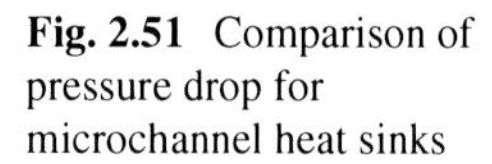
Fig. 2.51 Comparison of pressure drop for microchannel heat sinks

2.5.3 *Parametric Study*

There are a few key design parameters that greatly influence the heat transfer and pressure drop performance in oblique finned microchannels. As these parameters vary, the overall fin layout will change. This change has an enormous effect on the flow field and consequently it affects the performance of the heat sink. Therefore, a parametric study is essential to explore the outcome of varying these parameters. The two critical design parameters identified for the parametric study are oblique angle and oblique fin pitch. A total of seven configurations with three oblique angles (15°, 27° and 45°) coupling with three fin pitches (400, 800 and 1,500 μm) are fabricated on silicon-based microchannel heat sink for the performance evaluation. The benchmarking is based on two conventional microchannel configurations, which have 100 and 200 μm nominal channel width, respectively. Detailed dimensions of various microchannel heat sinks are tabulated in Tables 2.5 and 2.6. Microchannel heat sinks with a similar channel width, fin pitch but different oblique angles are grouped together to compare the effect of oblique angle variation, and it is vice versa when evaluating the effect of fin pitch variation.

While the combination of a different oblique angle and pitch can result in a wholly new enhanced microchannel heat sink layout, it also redistributes the fin and un-finned surface areas available for convective heat transfer. Table 2.7 compares the fin, un-finned and total surface areas available for convective heat transfer of different microchannel configurations. From Table 2.7, it is clear that the introduction of oblique fins and channels in microchannel heat sink increases the total heat transfer area by 5–30 % with smaller fin pitch and oblique angle leading to larger surface areas.

Table 2.5 Dimensional details of microchannel heat sinks with 100 μm channel width

Characteristic	Conventional microchannel #1 (Con-100)	Enhanced microchannel #1 (27°—NP′)	Enhanced microchannel #2 (45°—NP′)
Material	Silicon		
Footprint, width × length (mm^2)	12.7 × 12.7		
Number of fin row, n	61		
Main channel width, w_c (μm)	115	113	110
Fin width, w_w (μm)	85	87	91
Channel depth, H (μm)	387	379	351
Aspect ratio, α	3.37	3.35	3.19
Oblique channel width, w_{ob} (μm)	–	49	48
Fin pitch, p (μm)	–	405	400
Fin length, l (μm)	–	292	332
Oblique angel, θ (°)	–	26.3	45.0

—*NP′* denotes nominal pitch (~400 μm) for enhanced microchannel heat sink with 100 μm channel width

2.5.3.1 The Effect of Oblique Angle Variation

The effect of a change in oblique angle to heat transfer, and pressure drop performance, is diagnosed with three groups of microchannel heat sinks. The first group is three microchannel heat sinks with 100 μm nominal channel width, which consists of a conventional straight channel heat sink (Con-100) as benchmark and two enhanced heat sinks (~400 μm nominal fin pitch) with 27° and 45° oblique angles, respectively (27°—NP′ and 45°—NP′). The experimental results of both heat transfer and pressure drop are presented in Fig. 2.52. Figure 2.52a compares the average Nusselt number of three microchannel heat sinks. It is observed that the average Nusselt numbers of Con-100 and 45°—NP′ almost overlap each other, while that of 27°—NP′ is significantly higher. This suggests that 45°—NP′ is not as effective as 27°—NP′ in augmenting heat transfer performance. In contrast, the heat transfer augmentation for 27°—NP′ rises with Reynolds number. On the other hand, the pressure drop across Con-100 and 45°—NP′ overlaps each other as they did in the Nusselt number plot, as shown in Fig. 2.52b. The 27°—NP′ also displays a pressure drop that is comparable to the other two for Reynolds numbers lower than 500, signalling that the heat transfer enhancement can be achieved with negligible pressure drop penalty.

The second group of microchannel heat sinks for comparison has the same features as the first group, with exception that they are 200 μm in nominal channel width. In this group, conventional microchannel (Con-200) is adopted as the baseline for comparison with two enhanced microchannel heat sinks (~800 μm nominal fin pitch) with 27° and 45° oblique angles, respectively (27°—NP and 45°—NP). A similar trend is observed from the heat transfer and pressure drop performance of

Table 2.6 Dimensional details of microchannel heat sinks with 200 μm channel width

Characteristic	Conventional microchannel #2 (Con-200)	Enhanced microchannel #3 (27°—NP)	Enhanced microchannel #4 (45°—NP)	Enhanced microchannel #5 (15°—LP)	Enhanced microchannel #6 (27°—LP)	Enhanced microchannel #7 (45°—SP)
Material	Silicon					
Footprint, width × length (mm^2)	12.7 × 12.7					
Number of fin row, n	30					
Main channel width, w_c (μm)	205	205	203	205	206	204
Fin width, w_w (μm)	195	195	197	194	194	196
Channel depth, H (μm)	450	418	445	455	405	449
Aspect ratio, α	2.20	2.04	2.19	2.17	1.97	2.20
Oblique channel width, w_{ob} (μm)	–	104	103	101	103	103
Fin pitch, p (μm)	–	783	800	1,502	1,536	400
Fin length, l (μm)	–	550	653	1,104	1,301	253
Oblique angel, θ (°)	–	27.1	45.0	14.9	26.3	45.0

—*NP* denotes nominal pitch (~800 μm) for enhanced microchannel heat sink with 200 μm channel width
—*LP* denotes large pitch (~1,500 μm) for enhanced microchannel heat sink with 200 μm channel width
—*SP* denotes small pitch (~400 μm) for enhanced microchannel heat sink with 200 μm channel width

Table 2.7 Comparison of convective heat transfer areas for microchannel heat sinks

Configuration	Oblique angle (°)	Fin pitch (μm)	Oblique fin perimeter (μm)	Fin area (mm²)	Un-finned area (mm²)	Total heat transfer area (mm²)
Con-100	–	–	–	609.5	90.6	700.1
27°—NP′	26.3	405	976.6	718.7	107.5	826.2
45°—NP′	45.0	400	919.8	635.6	98.1	733.7
Con-200	–	–	–	354.0	80.8	434.8
27°—NP	27.1	783	1,956.9	409.1	102.8	511.9
45°—NP	45.0	800	1,864.8	406.4	93.6	500.0
15°—LP	14.9	1,502	3,718.2	441.6	100.5	542.1
27°—LP	26.3	1,536	3,476.4	359.1	92.4	451.5
45°—SP	45.0	400	1,059.3	464.5	107.8	572.3

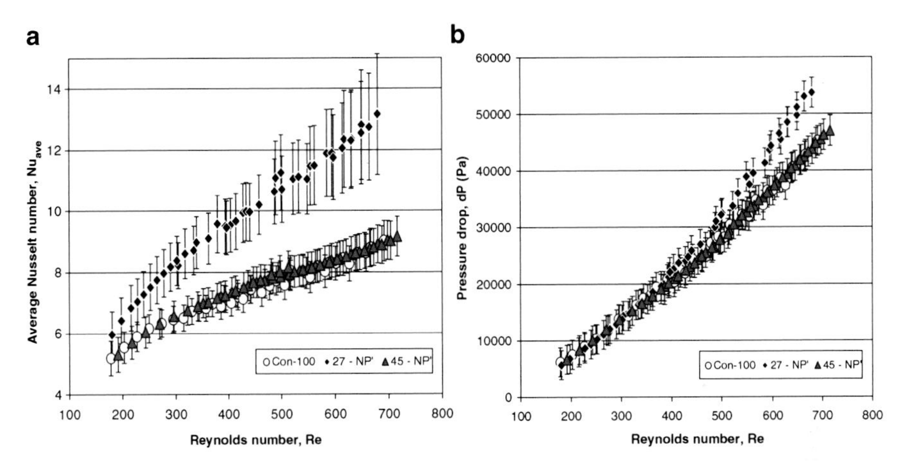

Fig. 2.52 Comparison of (**a**) average Nusselt number, (**b**) pressure drop for microchannel heat sinks with different oblique angles (100 μm nominal channel width and 400 μm nominal fin pitch)

this group, as displayed in Fig. 2.53 in comparison to the first group, which has 100 μm nominal channel width. Both Con-200 and 45°—NP result in comparable average Nusselt numbers and pressure drop for Reynolds number up to 1,000, while that of 27°—NP is consistently higher. This again indicates that smaller oblique angle is better for heat transfer enhancement. The gradient for 45°—NP in the average Nusselt number increases beyond Re 1,000, suggesting that this configuration can be more effective in heat transfer for higher Reynolds number.

From the assessment employing multiple microchannel groups, it is consistently demonstrated that a smaller oblique angle contributes to higher heat transfer performance, with a generally higher pressure drop. The smaller oblique angle results in lesser flow resistance than the larger oblique angle. Similar to liquid flow through a bend, loss coefficient increases for sharper bend as area of flow separation becomes extensive [23]. Thus, a smaller oblique angle helps to translate the diffusive oblique

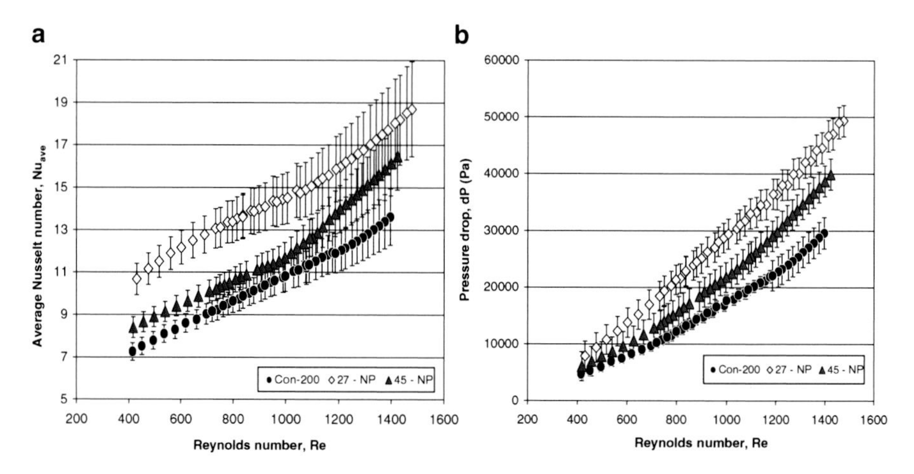

Fig. 2.53 Comparison of (**a**) average Nusselt number, (**b**) pressure drop for microchannel heat sinks with different oblique angles (200 μm nominal channel width and 800 μm nominal fin pitch)

channel into a smoother and smaller flow area expansion from the main channel. This can help to reduce the boundary layer separation that occurs in the oblique channel and increase the secondary flow generation. However, one should be aware that for a smaller oblique angle with similar fin pitch, oblique fins will become thinner, and this might compromise the structural integrity. In fact, the author has struggled to fabricate the enhanced microchannel heat sinks with 15° oblique angle through mechanical wafer cutting.

2.5.3.2 The Effect of Oblique Fin Pitch Variation

The impact of oblique fin pitch variation to the enhanced microchannel performance is evaluated in two groups of microchannel heat sinks, with 200 μm nominal channel width, by varying the oblique fin pitch while other parameters including oblique angle are fixed. The first group of microchannel heat sinks includes a conventional microchannel (Con-200) and two enhanced microchannel heat sinks (both have 27° oblique angle) with 800 and 1,500 μm fin pitches, respectively (27°—NP and 27°—LP). The second group of microchannel heat sinks for comparison comprises of conventional microchannel (Con-200) and two enhanced microchannel heat sink (both have 45° oblique angle), with 400 and 800 μm fin pitch (45°—SP and 45°—NP). The experimental heat transfer and pressure drop findings of the first group is presented in Fig. 2.54. The average Nusselt number is the highest for 27°—NP followed by 27°—LP while that of Con-200 is the lowest. It also suggests that the maximum temperature is lower than conventional microchannel. On the other hand, the pressure drop for 27°—NP is the highest while both 27°—LP and Con-200 display comparable pressure drop. This indicates that shorter fin pitch is beneficial for the heat transfer but incurs some pressure drop penalty.

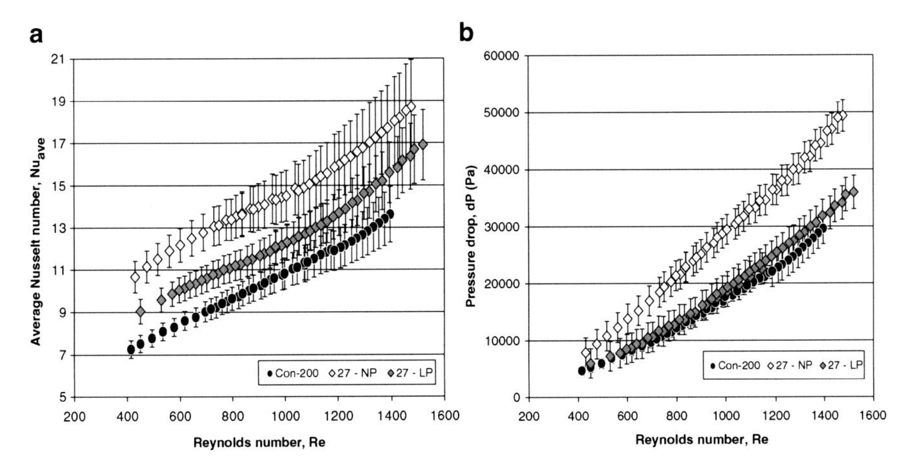

Fig. 2.54 Comparison of (**a**) average Nusselt number, (**b**) pressure drop for microchannel heat sinks (200 μm nominal channel width and 27° oblique angle) with different fin pitches

2.5.3.3 Overall Performance Comparison

Besides evaluating the heat transfer of microchannel heat sinks to the Reynolds number, the total thermal resistance of the individual heat sink is always compared between each other under the constant pressure drop and, also, pump power constraint, in the selection process to determine the best performing heat sink that meets the operating constraints. Figure 2.55 demonstrates the comparison of total thermal resistances of various microchannel heat sinks under the constraint of pressure drop. Among all the heat sinks tested, the enhanced microchannel with 100 μm channel width and 27° oblique channel (27°—NP′) achieves the lowest total thermal resistance at the lowest pressure drop. This observation demonstrates the effectiveness of the enhanced microchannel through the integration of smaller channel size (hydraulic diameter) that has large surface area-to-volume ratio with short sectional oblique fins. For instance, the pressure drop that is required to achieve 0.119 °C/W employing conventional microchannel (Con-100) can be reduced by ~55 % (from 45 to 25 kPa) by adopting the enhanced microchannel (27°—NP′).

A similar trend is observed for microchannel heat sinks with 200 μm nominal channel width. 45°—SP and 15°—LP are the two microchannel configurations that record the lowest total thermal resistance, in comparison to other microchannel heat sinks with 200 μm nominal channel width. This finding also points out the effectiveness of a smaller fin pitch and a smaller oblique angle in enhancing heat transfer performance. More importantly, these configurations perform better than conventional microchannel heat sink with 100 μm nominal channel width at the similar pressure drop. By incorporating sectional oblique fins, both channel width and fin width of a microchannel heat sink can be relaxed, allowing a much simpler, and economical, fabrication process.

On the other hand, Fig. 2.56 compares the total thermal resistances to the pump power requirement. In this context, pump power is computed as the product of volu-

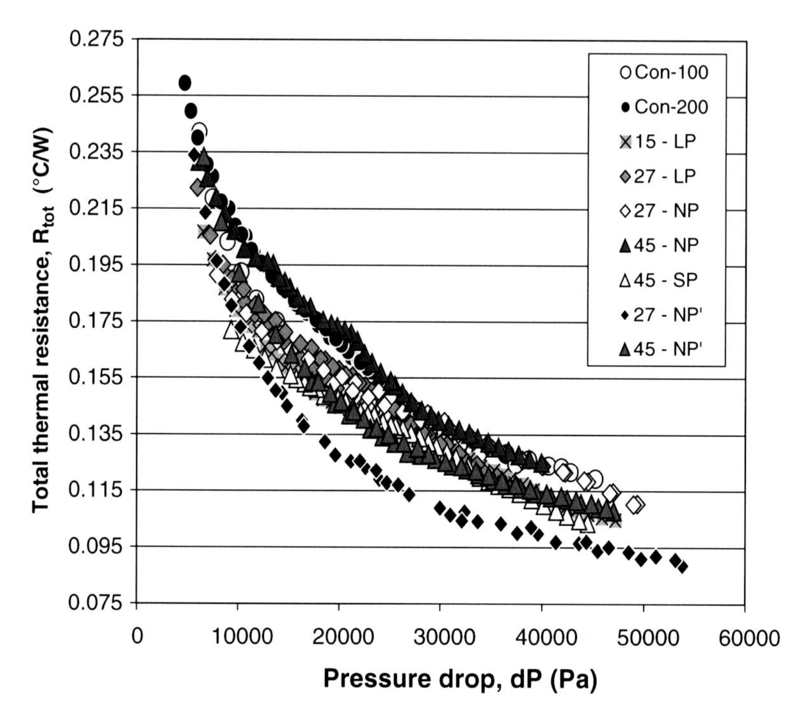

Fig. 2.55 Comparison of total thermal resistance to the pressure drop across microchannel heat sinks

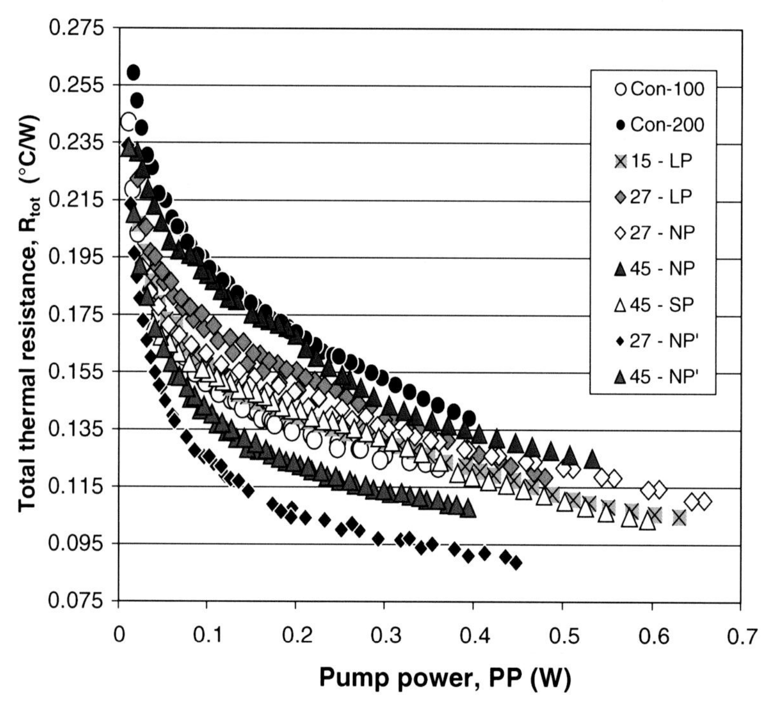

Fig. 2.56 Comparison of total thermal resistance to the pump power required for microchannel heat sinks

metric flow rate and pressure drop across the microchannel heat sink. Through this comparison, it is clear that enhanced microchannel heat sinks with 100 μm nominal channel width dominate those with 200 μm nominal channel width as the latter require much higher volumetric flow rate to achieve a comparable heat transfer performance. However, it is found that some enhanced microchannel heat sinks with 200 μm nominal channel width, for instance, 45°—SP and 15°—LP, can perform as well as conventional microchannel heat sink with 100 μm nominal channel width. This parametric study provides alternatives for the further performance improvement of larger size channel, should the fabrication present a constraint on the minimum channel size. In summary, the heat transfer of microchannel heat sinks can be ranked as follows under constant pressure drop or pump power constraint:

27°—NP′ > 45°—NP′ > Con-100, 45°—SP and **15°—LP > 27°—NP > 27°—LP > 45°—NP > Con-200**

2.6 Hotspot Mitigation with Oblique Fin Microchannel Heat Sink

It has been demonstrated in the previous sections that enhanced microchannel with oblique fins can be employed to improve the heat transfer performance of microchannel heat sink under uniform heating conditions. However, it is desirable to propose a thermal management technique that can cater for highly non-uniform heat flux dissipation, which has become the trend of advanced electronics. This section examines the effect of selectively varying the oblique fin pitch in the microchannel heat sink. Numerical simulation shows that denser oblique fin cluster promotes higher occurrence of boundary layer redevelopment and secondary flow generation, resulting in local heat transfer enhancement, without the penalty of increased pressure drop. Subsequently, experimental investigation with two hotspot conditions is conducted to demonstrate and validate the feasibility of this concept.

2.6.1 Hotspot Mitigation Concept with Oblique Fins

There is an urgent requirement to introduce a new thermal management technique to cope with the hotspots on electronic devices, as the conventional cooling schemes, which are designed for uniform heat flux dissipation, are not effective in cooling hotspots [31]. Thus, this section explores the feasibility of adapting enhanced microchannel heat sink, with oblique fin, for electronic hotspot mitigation [32]. It would be interesting to examine the effect of selectively varying the oblique fin pitch based on the local heat flux level to control the occurrence of boundary layer redevelopment and secondary flow generation, which in turn tailor the required local heat transfer performance. This idea is inspired by Sui et al. [28], who proposed to vary the waviness of wavy channel based on local heat flux level to tailor the chaotic advection in the wavy channel.

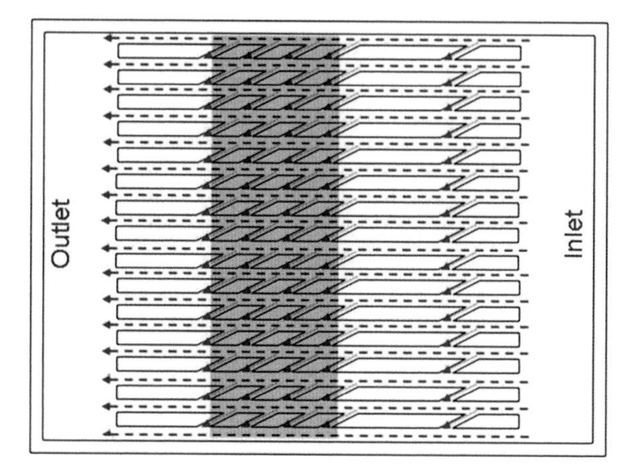

Fig. 2.57 Plan view of variable pitch oblique finned microchannel heat sink with hotspot (*red colour highlighted*)

Shorter fin pitch leads to closely packed oblique fins and channels, where thermal boundary layer redevelopment and secondary flow generation would occur at higher frequency. Consequently, local heat transfer performance can be greatly enhanced. In contrast, longer fin pitch reduces the number of oblique fin, as well as the associated heat transfer enhancement. The combination of these features turns out to be ideal for hotspot thermal management, where denser oblique fin cluster provides effective heat removal for the extreme high heat flux, while sparser oblique fin cluster dissipates the background heat flux without overcooling the chip. A potential embodiment of variable pitch oblique finned microchannel heat sink is shown in Fig. 2.57. In addition, variable pitch oblique finned microchannel heat sink is expected to be effective especially for multiple hotspot configuration, where the denser fin clusters can be positioned directly on top of the hotspots without affecting the oblique fin characteristics in the neighbouring cluster.

2.6.2 Experimental Set-Up and Procedure

Two hotspot scenarios are considered in this experimental investigation; the first has a single hotspot at the centre of the thermal test chip, while the second has multiple hotspots over the entire chip. Three different heat sink designs, namely, conventional, constant pitch oblique finned and variable pitch oblique finned microchannel heat sink, are evaluated for their heat dissipating performance.

A similar experimental set-up that was employed in the study for enhanced microchannel heat sink with silicon thermal test chips in Sect. 2.4.1 is adopted for the investigation of electronic hotspot mitigation. An additional DC power supply and the associate current (power) measuring components (shunt resistor, voltage meter, etc.) are required to enable different level of heat supply into thermal test chip, to emulate the hotspot scenario. In these experiments, thermal test chips are supplied with two different heat fluxes: a background heat flux at the lower magnitude for most of the chip area and a significantly higher hotspot heat flux for small area of the chip area.

Two hotspot scenarios are considered in this experimental investigation; the first has a single hotspot at the centre of the thermal test chip, and the second has multiple hotspots over the entire chip. The same conventional microchannel and enhanced microchannel, with uniform fin pitch, are used for both experimental conditions. Separate enhanced microchannels with variable fin pitch are fabricated to tackle different hotspot conditions with local heat transfer performance tailoring. A cluster of oblique fins, with finer fin pitch, are placed on top of each hotspot with the intention of increasing both heat transfer area and cooling capability. The design for enhanced microchannel heat sinks with variable fin pitch is illustrated in Fig. 2.58a, b, where the red patches represent the location of the hotspot on the thermal test chip and the corresponding fin structure. Figure 2.59a, b shows the actual test vehicles of enhanced microchannel heat sinks with variable fin pitch.

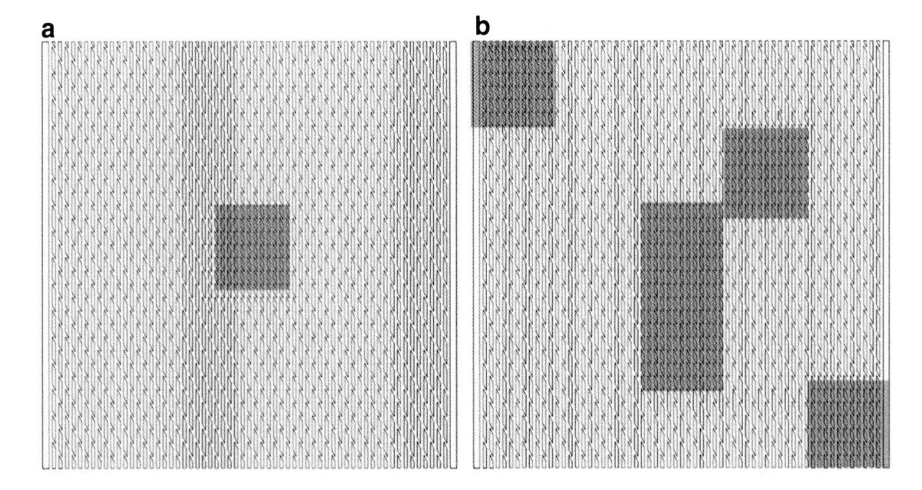

Fig. 2.58 Heat sink design for enhanced microchannel with variable fin pitch for (**a**) single hotspot condition and (**b**) multiple hotspot condition [32]

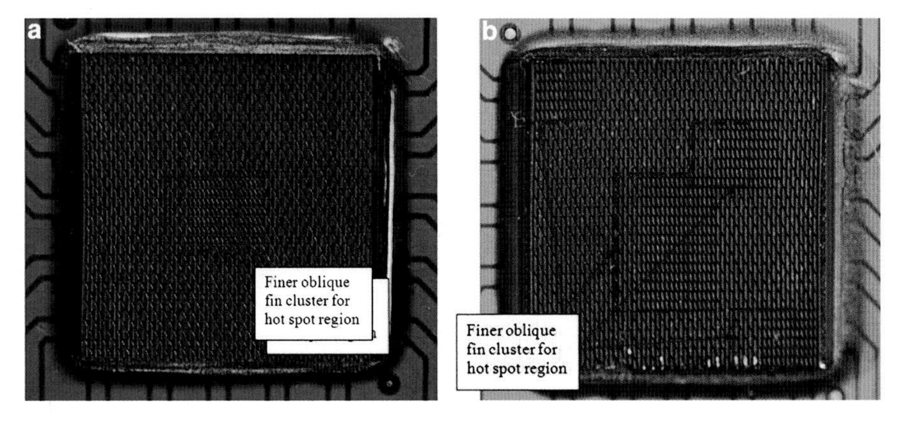

Fig. 2.59 Actual test vehicle for enhanced microchannel heat sink with variable fin pitch for (**a**) single hotspot and (**b**) multiple hotspots [32]

All the heat sinks are fabricated using laser machining, as the conventional fabrication methods such as wire cut and saw dicing are not able to produce oblique channel/fin at different pitch within a heat sink. Some important process parameters or settings for the laser machining are documented in Table 2.8.

The critical geometries are then measured with 3D profiler, and the data are tabulated in Table 2.9. Each characteristic is computed as an average value of 6 measurements across the heat sink. Furthermore, the fin channel profile in Fig. 2.60 shows that the channels are tapered at the bottom, as a result of the naturally pointed profile of a laser beam. Triangular channels are thus created instead of the rectangular channels achieved for the previous test vehicles. For a fair comparison, main channel width, fin width, oblique channel width, oblique channel angle and aspect ratio of different heat sinks are kept at a comparable range. The average roughness of the machined surfaces is measured at 0.91 μm.

Table 2.8 Process parameters for laser machining microchannel heat sinks

Characteristic	
Laser type	Nd: YAG
Laser power	150 μJ
Laser frequency	40 kHz
Wavelength	355 nm
Scanning lens	F-Theta
Travel speed	300 mm/s
Number of pass	40

Table 2.9 Dimensional details of test pieces used in the experiment [32]

Characteristic	Conventional microchannel	Enhanced microchannel (uniform fin pitch)	Enhanced microchannel (variable fin pitch—single hotspot)	Enhanced microchannel (variable fin pitch—multiple hotspots)
Material	Silicon			
Footprint, width × length (mm^2)	12.7 × 12.7			
Number of fin row, N	61			
Main channel width, w_c (μm)	102	100	105	94
Fin width, w_w (μm)	97	96	101	101
Channel depth, H (μm)	306	300	312	328
Aspect ratio, α	3.0	3.0	2.97	3.49
Oblique channel width, w_{ob} (μm)	–	65	63	59
Fin pitch, p_L (μm)	–	799	782	762
Fin length, l_L (μm)	–	636	650	642
Fin pitch, p_s (μm)	–	–	388	386
Fin length, l_s (μm)	–	–	252	271
Oblique angel, θ (°)	–	26.7	26.8	27.4

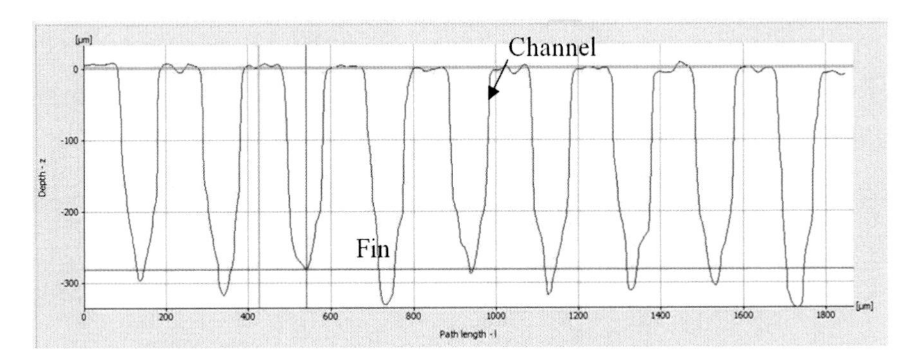

Fig. 2.60 Fin channel profile captured with 3D profiler [32]

A similar experimental procedure, as stated in Sect. 2.4.1, is adopted in this study. Total coolant flow rate across the heat sink is fixed at 200 mL/min for all the experimental runs, which corresponds to the Reynolds number of ~600. The background heat flux of the thermal test chip is set as 85 W/cm^2. On the other hand, hotspot heat flux for the single hotspot scenario is tuned to 400 W/cm^2, while that of the multiple hotspot scenario is at 300 W/cm^2.

The significant difference in the heat flux level between hotspots and remaining chip area would result in severe heat flux redistribution. 3D conduction occurs for the discrete heat fluxes applied at the bottom of chip towards the channel wall of the liquid passages. Unfortunately, local heat flux distribution at the channel wall is not measurable in the current experiment. Without this critical information, computation of local Nusselt number would be baseless. Due to this limitation, the discussion of the results is restricted to overall temperature contour, maximum temperature, temperature gradient and pressure drop across the heat sinks. Pressure drop across the microchannel heat sinks is reduced according to the method stated in Sect. 2.2.1.

2.6.3 *Hotspot Scenarios*

In this section, the experimental results of heat sink performance under single hotspot scenario are first discussed, followed by the findings of heat sink under multiple hotspot condition.

2.6.3.1 Single Hotspot Scenario

Comparisons between the maximum temperatures for the three heat sinks in Fig. 2.61 show that both enhanced microchannel heat sinks reduce the maximum chip temperature significantly. The maximum temperature of thermal test chip with conventional microchannel heat sink is at 77.6 °C, while that of the enhanced

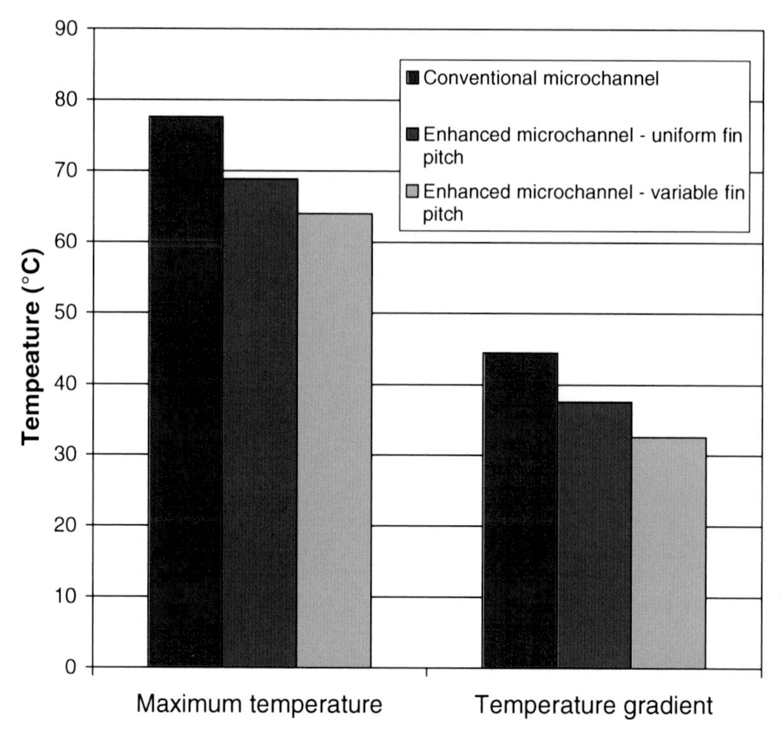

Fig. 2.61 Comparison of maximum chip temperature and temperature gradient for microchannel heat sinks [32]

microchannel with uniform fin pitch and variable fin pitch is reduced to 68.9 °C (8.7 °C reduction) and 64.0 °C (13.6 °C reduction), respectively. This shows that effective hotspot suppression can be achieved by employing sectional oblique fins. In the uniform oblique fin pitch configuration, the combined effect of thermal boundary layer redevelopment and secondary flow generation results in greater convective heat transfer performance, which lowers the chip temperature. On the other hand, variable oblique fin pitch configuration provides the opportunity to further improve the heat transfer performance locally on the hotspot. In addition to increasing the heat transfer area, clustering oblique fins with finer fin pitch leads to higher occurrence of thermal boundary layer redevelopment and secondary flow generation, which further elevates the heat transfer performance of the region. As a result, the maximum temperature of the hotspot is further lowered.

A similar trend is observed for the temperature gradient of thermal test chips. This is defined as the difference between local maximum and minimum temperature of the test chip, chip temperature gradient employing conventional microchannel heat sink, enhanced microchannel heat sink with uniform fin pitch and enhanced microchannel heat sink with variable fin pitch, 44.5, 37.6 and 32.6 °C, respectively. The reduction in temperature gradient occurs as a result of significantly lower maximum chip temperature for the enhanced microchannel heat sinks. By achieving the lowest maximum chip temperature and temperature gradient among three cooling techniques, the enhanced microchannel heat sink with variable fin pitch demonstrates its capability as an effective solution in electronic hotspot mitigation by flexibly adapting to the heating condition.

Figure 2.62a–c illustrates the detailed chip temperature distribution for each silicon chip based on the temperature measured from all 25 temperature sensors (in 5×5 grid array) within the thermal test chip. Generally, hotspots with maximum temperatures occur at the centre of the thermal test chip, where the hotspot heat flux (400 W/cm^2) is significantly higher than the background heat flux (85 W/cm^2). As discussed, conventional microchannel heat sink results in the highest hotspot temperature followed by enhanced microchannel with uniform fin pitch and enhanced microchannel with variable fin pitch. Beyond the hotspot, chip temperature is reduced but the region located downstream to the hotspot tends to have higher temperature compared to its surrounding at the same axial distance due to the larger sensible heat gain as coolant travels past the hotspot. In addition, it is noticed that the temperature contour is rather "squarish" in shape, possibly due to the limited 25 data points available in square grid array.

At a similar background heat flux at 85 W/cm^2, the enhanced microchannel heat sink with uniform fin pitch lowers the temperature of the entire chip by at least 5 °C compared with conventional microchannels. The heat transfer augmentation, brought forth by the oblique fins/channels, is uniform across the test chip. A temperature spike at the hotspot is also milder for enhanced microchannel, where the temperature difference between the hotspot and the surroundings is reduced. As for the enhanced microchannel with variable fin pitch, chip temperature for regions under the background heat flux is maintained at an almost identical value of the enhanced microchannel with uniform fin pitch, while the maximum temperature at the hotspot is further reduced leading to a much more uniform temperature contour. Such design not only suppresses electronic hotspot but also avoids the overcooling of remaining chip area, which is the critical criterion of a hotspot mitigation scheme [31]. If oblique fins with finer fin pitch are laid across the entire heat sink, it is highly possible that maximum hotspot temperature would remain the same while both temperature gradient and pressure drop across the heat sink will be increased.

Examining pressure drop across the microchannel heat sinks shows that pressure drop for conventional microchannel is the lowest among the threes. Although the pressure drop for both enhanced microchannel is higher, the difference is not significant. From Fig. 2.63, the pressure drop for both enhanced microchannel heat sinks is about 10–20 % higher than that of the conventional microchannel. This shows that by adopting an effective thermal management scheme, heat transfer performance can be augmented significantly, with affordable pressure drop penalty.

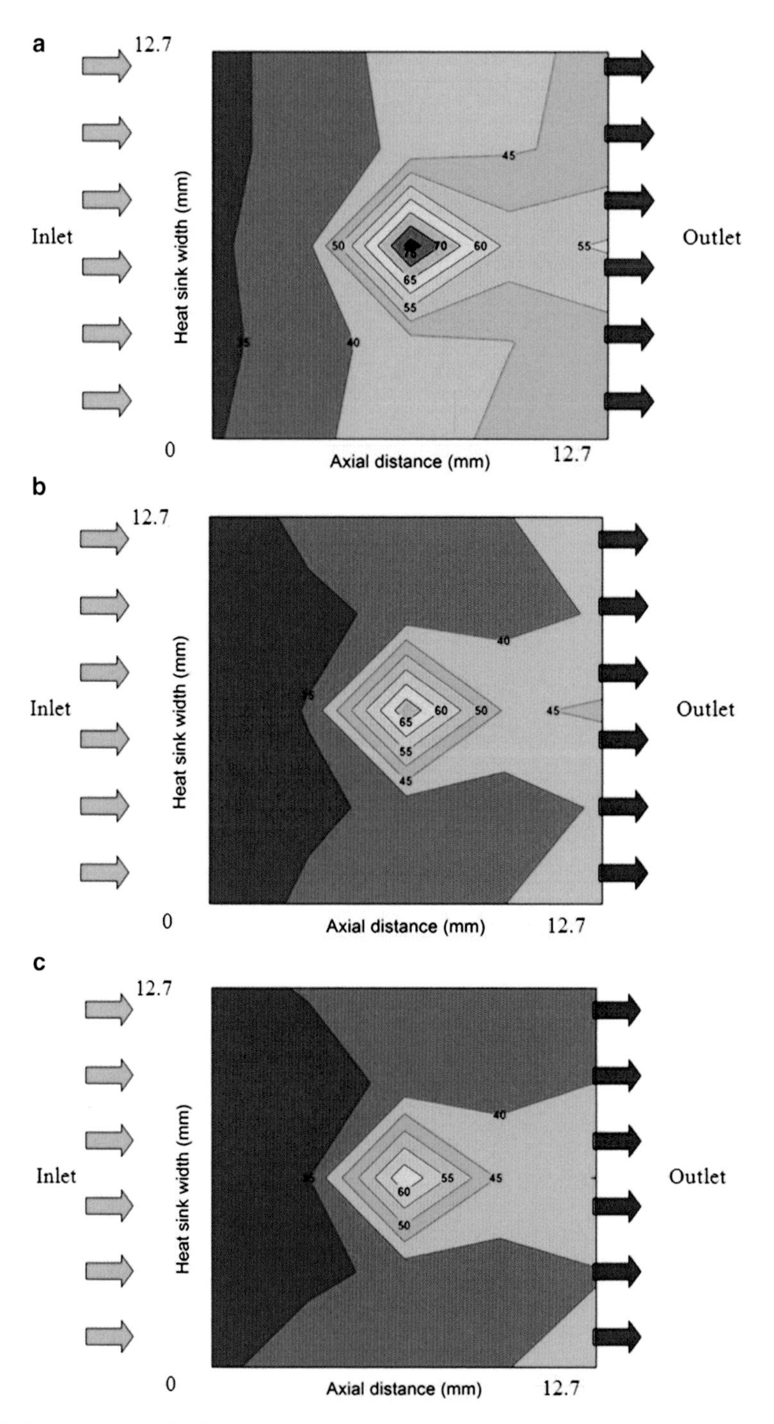

Fig. 2.62 Temperature distribution for single hotspot scenario for (**a**) conventional microchannel, (**b**) enhanced microchannel with uniform fin pitch, and (**c**) enhanced microchannel with variable fin pitch [32]

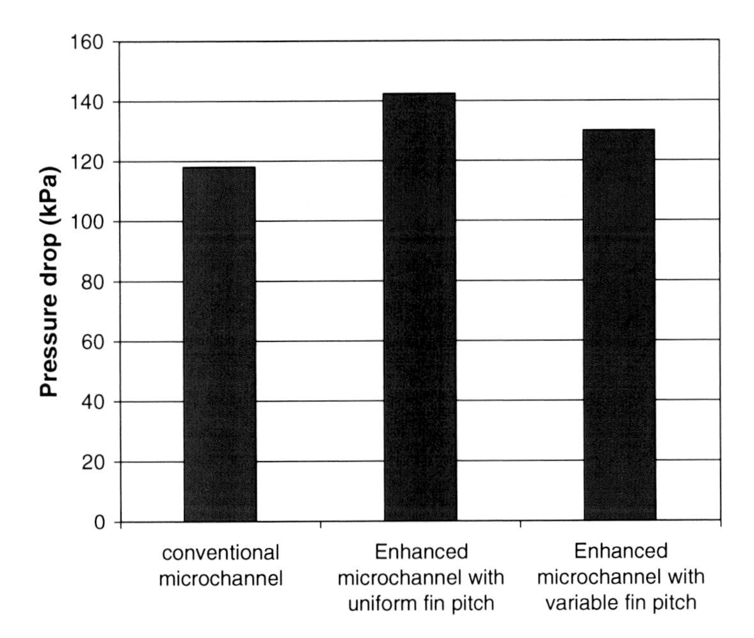

Fig. 2.63 Comparison of pressure drop across microchannel heat sinks [32]

2.6.3.2 Multiple Hotspot Scenario

A total of five hotspots, each dissipating 300 W/cm² heat flux, are emulated on the thermal test chip, with a background heat flux of 85 W/cm². The hotspots are positioned on the chip as illustrated in Fig. 2.59b, with two of hotspots placed adjacent to each other. As a result, the temperature contour of the thermal test chip becomes more complex compared to the single hotspot scenario. Figure 2.64a illustrates the temperature distribution of the chip with conventional microchannel heat sink under the multiple hotspot scenario. The occurrence of multiple hotspots in the chip changes the chip temperature contour completely. Despite showing a general trend of progressive temperature increment in the axial direction, temperature spike at each hotspot is clearly observed. The maximum hotspot temperature can be as high as 75 °C, and the temperature spike at the hotspots is generally 20–25 °C hotter than the surrounding area that is just a few millimetres away. Such a high temperature gradient within a short distance in electronic chip is highly undesirable as it exerts additional thermal stress to the chip. This situation seems unavoidable as multiple functional modules with different level of power consumption being packed within a small chip as the industry move towards "more-than-Moore" technologies [33].

Nevertheless, this condition can be improved by adopting the better thermal management scheme. Figure 2.64b illustrates the temperature distribution of the

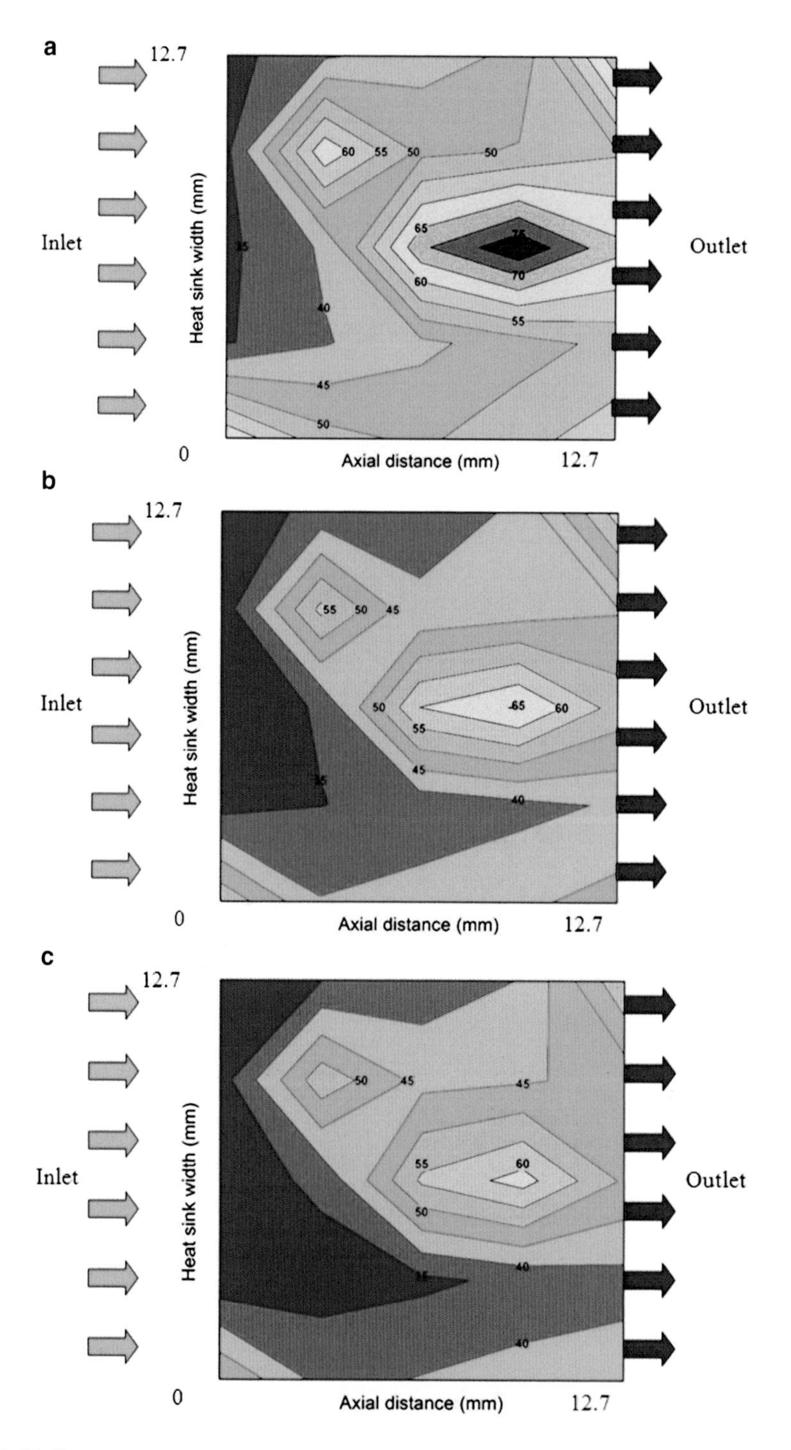

Fig. 2.64 Temperature distribution for multiple hotspot scenario for (**a**) conventional microchannel, (**b**) enhanced microchannel with uniform fin pitch, and (**c**) enhanced microchannel with variable fin pitch [32]

thermal test chip employing the obliquely finned microchannel heat sink with uniform fin pitch. Significant drop in temperature is observed across the entire chip, especially the hotspot region. A comparison between the temperature distributions for both heat sinks also shows that the enhanced microchannel heat sink is more effective in relieving the hotspot situation at the downstream of the heat sink. The magnitude of temperature reduction for hotspots at the upstream of the heat sink is ~5 °C, while that for the downstream of the heat sink can be larger than 10 °C. This can be attributed to the unique feature of the enhanced microchannel heat sink in maintaining a uniform heat transfer performance in the axial direction. While heat transfer performance deteriorates in the axial performance for a conventional microchannel, the enhanced microchannel exhibits the higher, yet uniform, heat transfer coefficient owing to the thermal boundary layer redevelopment and secondary flow generation that occur throughout the heat sink. This combined effect creates a constantly renewed flow field that is beneficial, especially for easing off the hotspots downstream of the heat sink.

Figure 2.64c, on the other hand, displays the temperature contour of a chip employing the enhanced microchannel heat sink with variable pitch. The temperatures are generally similar to that of the enhanced microchannel with uniform fin pitch except the hotspots, where the temperatures are further lowered. By adopting finer fin pitch for hotspots and coarser fin pitch for other regions with lower background heat flux, the heat sink performance can be tailored based on the heat flux level. Thus, the region with the higher heat flux dissipation will always have the higher heat removal capability and lower hotspot temperature, regardless of its location on the chip. As a result, a much more uniform temperature distribution can be created across the entire chip with this performance tailoring technique, rendering it as an effective thermal management scheme for electronics with hotspots.

For a hotspot mitigation scheme to be successful, the hotspot suppressing feature must be placed close to the hotspot. Thus, it is most effective if variable pitch oblique finned microchannel heat sink is integrated onto the silicon chip. A separate heat sink on a device with hotspots will see its advantage of local heat transfer performance tailoring be neutralized by the spreading and contact resistances.

The comparison for the maximum temperature and temperature gradient of the chips with different heat sinks is shown in Figure 2.65, which clearly demonstrates the reduction in maximum temperature and temperature gradient of the chip from conventional microchannel to enhanced microchannel with uniform fin pitch and variable fin pitch.

Figure 2.66 displays the pressure drop across the microchannel heat sinks for the multiple hotspot scenario where similar trend is observed compared with the single hotspot experimental condition. Conventional microchannel heat sink records the lowest pressure drop among the three, while that of the enhanced microchannel heat sinks are 10–20 % higher.

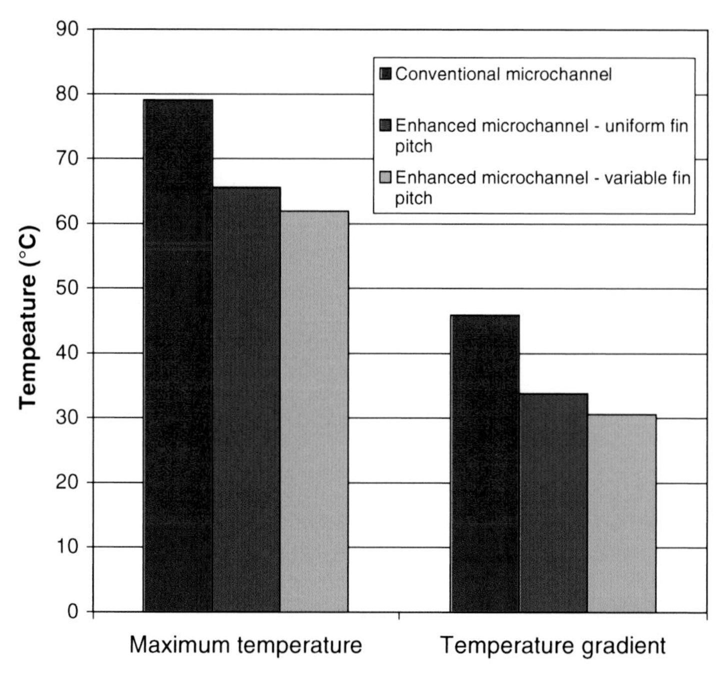

Fig. 2.65 Comparison of maximum chip temperature and temperature gradient for microchannel heat sinks [32]

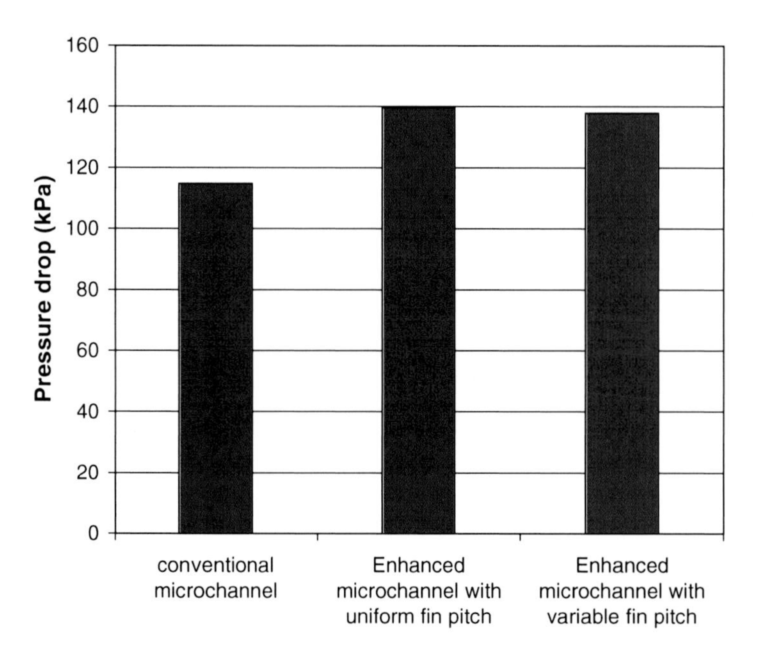

Fig. 2.66 Comparison of pressure drop across microchannel heat sinks for multiple hotspot scenario [32]

References

1. Steinke ME, Kandlikar SG (2004) Single-phase heat transfer enhancement techniques in microchannel and minichannel flows. In: Proceedings of the ASME 2nd international conference on microchannels and minichannels (ICMM 2004), Rochester, New York, USA, paper no-2328, pp 141–148, 17–19 June 2004
2. DeJong NC, Jacobi AM (2003) Localized flow and heat transfer interactions in louvered-fin arrays. Int J Heat Mass Transfer 46:443–455
3. Lee YJ, Lee PS, Chou SK (2013) Numerical study of fluid flow and heat transfer in the enhanced microchannel with oblique fins. J Heat Transfer 135:041901. doi:10.1115/1.4023029
4. Suga T, Aoki H (1991) Numerical study on heat transfer and pressure drop and pressure drop in multilouvered fins. In: Lloyd JR, Kurosake Y (eds) Proceedings of 1991 ASME/JSME joint thermal engineering conference, vol 4, ASME, New York, p 361–368
5. Davenport CJ (1980) Heat transfer and fluid flow in the louvered-fin heat exchanger. Ph.D. Thesis, Department of mechanical Engineering, Lanchester Polytechnic, UK
6. Aoki H, Shinagawa T, Suga K (1989) An experimental study of the local heat transfer characteristic in automotive louvered fins. Exp Therm Fluid Sci 2:293–300
7. FLUENT 6.3 user's guide (2006) Chapter 7: cell zone and boundary conditions, FLUENT Inc., Lebanon, NH
8. Li J, Peterson GP (2007) 3-Dimensional numerical optimization of silicon-based high performance parallel microchannel heat sink with liquid flow. Int J Heat Mass Tranfer 50:2895–2904. doi:10.1016/j.ijheatmasstransfer.2007.01.019, http://dx.doi.org
9. Ryu JH, Choi DH, Kim SJ (2002) Numerical optimization of the thermal performance of a microchannel heat sink. Int J Heat Mass Tranfer 45:2823–2827. doi:10.1016/S0017-9310(02)00006-6, http://dx.doi.org
10. Lee P-S, Garimella SV (2006) Thermally developing flow and heat transfer in rectangular microchannels of different aspect ratios. Int J Heat Mass Tranfer 49:3060–3067. doi:10.1016/j.ijheatmasstransfer.2006.02.011
11. Incropera FP (1999) Liquid cooling of electronic devices by single-phase convection. Wiley, New York, NY, pp 262–263
12. Lee P-S, Garimella SV, Liu D (2005) Investigation of heat transfer in rectangular microchannels. Int J Heat Mass Tranfer 48:1688–1704. doi:10.1016/j.ijheatmasstransfer.2004.11.019
13. Douglas JF, Gasiorek JM, Swaffield JA, Jack LB (2005) Fluid mechanics, 5th edn. Pearson, Harlow, UK, p 389
14. Alharbi AY, Pence DV, Cullion RN (2003) Fluid flow through microscale fractal-like branching channel networks. J Fluids Eng 125:1051–1057
15. Senn SM, Poulikakos D (2004) Laminar mixing, heat transfer and pressure drop in tree-like microchannel nets and their application for thermal management in polymer electrolyte fuel cells. J Power Sources 130:178–191
16. Chandratilleke TT, Jagannatha D, Narayanaswamy R (2010) Heat transfer enhancement in microchannels with cross-flow synthetic jets. Int J Therm Sci 49:504–513
17. Sahnoun A, Webb RL (1992) Prediction for heat transfer and friction for the louver fin geometry. J Heat Transfer 114:893–900
18. Xu JL, Gan YH, Zhang DC, Li XH (2005) Microscale heat transfer enhancement using thermal boundary layer redeveloping concept. Int J Heat Mass Tranfer 48:1662–1674
19. Xu JL, Song YX, Zhang W, Zhang H, Gan YH (2008) Numerical simulations of interrupted and conventional microchannel heat sinks. Int J Heat Mass Transfer 51:5906–5917
20. Kim SJ, Kim D, Oh HH (2008) Comparison of fluid flow and thermal characteristics of plate-fin and pin-fin heat sinks subjects to a parallel flow. Heat Transfer Eng 29:169–177
21. Lee YJ (2014) Fluid flow and heat transfer investigations on enhanced microchannel heat sink using oblique fins with parametric study. Int J Heat Mass Transfer (submitted)
22. Mou N (2014) Numerical investigations on fluid flow and heat transfer characteristic in oblique fin microchannel array. Int J Heat Mass Transfer (submitted)

23. Lee YJ, Lee PS, Chou SK (2012) Enhanced thermal transport in microchannel using oblique fins. J Heat Transfer 134:101901. doi:10.1115/1.4006843
24. Qu W (2004) Transport phenomena in single-phase and two-phase micro-channel heat sinks. Ph.D. Thesis, School of Mechanical Engineering, Purdue University, West Lafayette
25. Todreas NE, Kazimu MS (1990) Nuclear system I. Hemisphere, New York, NY
26. Lyman AC, Stephan RA, Thole KA, Zhang LW, Memory SB (2002) Scaling of heat transfer coefficients along louvered fins. Exp Therm Fluid Sci 26:547–563
27. Rosaguti NR, Fletcher DF, Haynes BS (2006) Laminar flow and heat transfer in a periodic serpentine channel with semi-circular cross-section. Int J Heat Mass Tranfer 49:2912–2923
28. Sui Y, Teo CJ, Lee PS, Chew YT, Shu C (2010) Fluid flow and heat transfer in wavy micro-channels. Int J Heat Mass Tranfer 53:2760–2772
29. Joshi HM, Webb RL (1987) Heat transfer and friction in the offset strip-fin heat exchanger. Int J Heat Mass Tranfer 30:69–84
30. Kandlikar SG, Colin S, Peles Y, Garimella S, Pease RF, Brandner JJ, Tuckerman DB (2013) Heat transfer in microchannels: 2012 status and research needs. J Heat Transfer 135:091001. doi:10.1115/1.4024354
31. Bar-Cohen A, Arik M, Ohadi M (2006) Direct liquid cooling of high flux micro and nano electronic components. Proc IEEE 94:1549–1570
32. Lee YJ, Lee PS, Chou SK (2013) Hotspot mitigating with obliquely finned microchannel heat sink: an experimental study. IEEE Trans Compon Packag Manuf Tech 3:1332–1341. doi:10.1109/TCPMT.2013.2244164
33. Zhang GQ, Graef M, Roosmalen FV (2006) The rationale and paradigm of "More than Moore". In: Proceedings of 56th electronic components and technology conference, San Diego. CA, pp 151–157. DOI: 10.1109/ECTC.2006.1645639

Chapter 3
Cylindrical Oblique Fin Minichannel Structure

Keywords Cylindrical oblique fin minichannel • Heat transfer • Pressure drop • Parametric study • Multiple correlations • Edge effect

3.1 Numerical Analysis of Cylindrical Oblique Fin Minichannel Heat Sink

In the previous chapter, it is noted that most studies on enhanced heat transfer focused mainly on flat heat source surfaces, and the existing study for cooling heat sources with cylindrical surface is rather scarce. Thus, it is worthwhile to propose an effective cooling solution for cylindrical heat source surfaces such as motors, generators, high-capacity battery and high-power LEDs. This section provides a description on the numerical simulation for both conventional and oblique fin minichannels and is divided into three main subsections. In Sect. 3.1.1, the details of cylindrical oblique fin minichannel geometry consideration and the CFD simulation approach are provided. In Sects. 3.1.2 and 3.1.3, simulation results and numerical analysis are discussed. The velocity and temperature profile, secondary flow distribution, entrance region effect, heat transfer and pressure drop characteristics are analysed in detail.

3.1.1 Cylindrical Oblique Fin Minichannel Geometry Consideration

A novel cylindrical oblique fin minichannel in the form of an enveloping jacket was proposed to fit over cylindrical heat sources such as motor, Li-ion battery, high-power laser rod, etc. The proposed design in Fig. 3.1 consists of two types of channel arrays, namely, main flow channels and oblique secondary channels [1]. The main flow channels are aligned with the axial direction, while the oblique secondary channels are branched out from the main flow channel at an oblique angle. The presence of the oblique secondary channels disrupts and reinitializes the boundary

Y. Fan et al., *Thermal Transport in Oblique Finned Micro/Minichannels*,
SpringerBriefs in Applied Sciences and Technology,
DOI 10.1007/978-3-319-09647-6_3

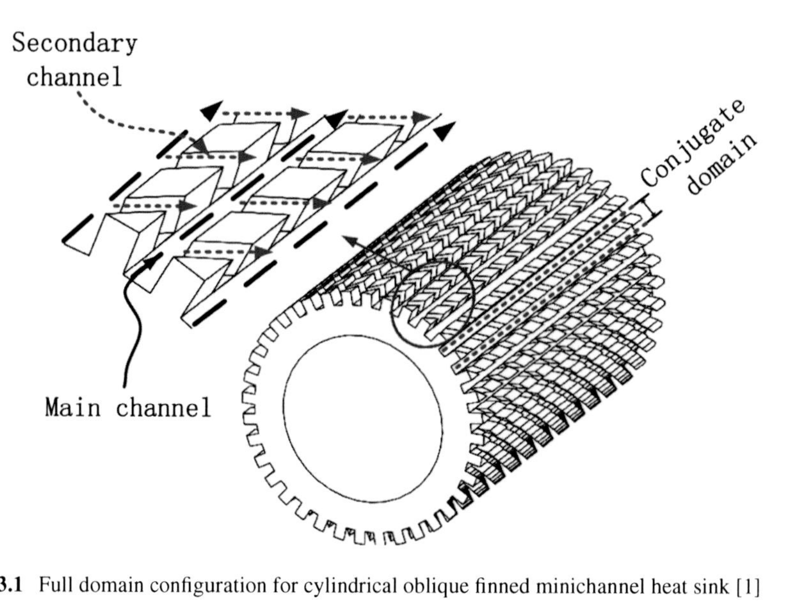

Fig. 3.1 Full domain configuration for cylindrical oblique finned minichannel heat sink [1]

layer development periodically. This results in the significant reduction of the boundary layer thickness and causes the flow to remain in the developing state unlike the conventional straight fin configuration. Furthermore, the length of oblique fin is much shorter compared to that of the conventional straight fin minichannel, which limits the boundary layer development effectively. Therefore, the wall temperature gradient, $\Delta T/\Delta y$, is consistently maintained at a higher value and higher heat transfer coefficient h can be achieved.

Lee et al. [2] employed sectional oblique fins to improve the flow mixing and enhance heat transfer performance on planar heat sink surface. It was reported that there was a slight deviation between numerical simulation results and experimental investigation, and it was attributed to the different flow behaviours between the side edge regions and the middle regions of the heat sink. Therefore, instead of utilizing a planar structure, this work proposes to use a cylindrical oblique fin minichannel structure with the intention to (1) better control the temperature of the cylindrical heat source, (2) eliminate the edge effect, (3) increase fluid–solid contact surface area, (4) generate uniform secondary flow on the minichannel surface, (5) disrupt thermal boundary layer development and (6) improve the flow mixing.

In this simulation, the geometrical dimensions and information for both novel cylindrical oblique fin minichannel and conventional minichannel are tabulated in Table 3.1. In order to facilitate a fair performance comparison, both heat sinks in the simulation study share the same aspect ratio, channel width, fin width and overall footprint. Apart from these common characteristics, the oblique angle that denotes the angle between main channel and secondary channel is set as ~30°, which referred to the planar oblique fin studies by Lee et al. [2], and it was also suggested in the louvre angle range (20°–45°) by Suga and Aoki [3] in their louvred fin heat exchangers studies.

Table 3.1 Dimension details for minichannel heat sinks

Characteristic	Conventional straight fin minichannel	Cylindrical oblique fin minichannel
Material	Copper	
Footprint dimension D, (mm)	18	
Fin length L, (mm)	65	
Number of rows	36	
Channel height H, (mm)	2	
Main channel bottom width (mm)	1.40	
Main channel top width (mm)	1.72	
Fin width (mm)	1.6	
Number of fins per row	–	11
Secondary channel gap (mm)	–	2
Oblique fin length (mm)	–	3.7
Oblique angle θ (°)	–	30

A numerical 3D conjugate heat transfer simulation is carried out with considerations on both heat convection in the channel and conduction in the copper substrate. Due to the similarity between this model and simplified model of planar heat sink design in Chap. 2, the CFD simulation approach, governing equation, grid independent study and boundary condition for present model can be referred to in Sect. 2.2.1.

3.1.2 *Flow Field Analysis*

In this section, detailed analysis of velocity and temperature profile, secondary flow distribution and entrance region effect is presented based on the numerical simulation results for conventional straight fin minichannel and cylindrical oblique fin minichannel.

3.1.2.1 Velocity and Temperature Profile

Simulation results reveal a clear flow field difference between the conventional straight fin minichannel and oblique fin minichannel. Figure 3.2 shows the velocity profile at mid-depth ($z=18$ mm) and mid-portion of the minichannel ($x=30$–35 mm) when the flow rate is 400 ml/min. From Fig. 3.2a, it can be seen that the high velocity gradient from the wall to the fluid core implies the hydraulic boundary layer is fully developed and maintained throughout the whole conventional straight fin minichannel. Nevertheless, in Fig. 3.2b, it is observed that the sectional oblique fin disrupts the velocity profile at each entrance of the downstream fin and causes the hydrodynamic boundary layer development to reinitialize at every downstream oblique fin.

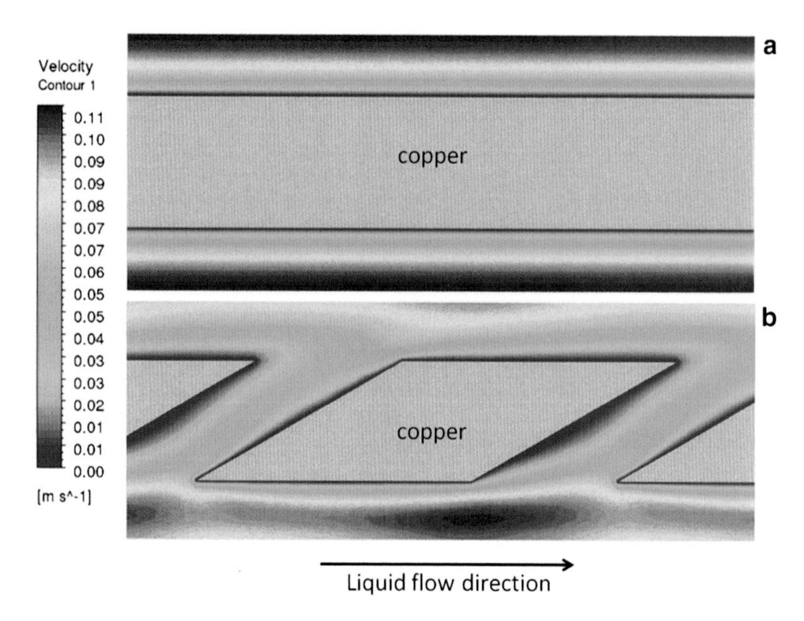

Fig. 3.2 Velocity contour for flow inside (**a**) conventional straight fin minichannel, (**b**) oblique fin minichannel [1]

This results in the boundary layer thickness reducing significantly in comparison with the conventional straight fin minichannel. Thus, the velocity profile is maintained in the developing region for this oblique fin structure.

Convective heat transfer takes place through both diffusion and advection. Heat is transported from copper surface into the fluid particle and propagates further into the fluid core. Due to the significant flow field difference, a large fluid temperature distinction is found between the conventional straight fin minichannel and oblique fin minichannel. In Fig. 3.3a, it can be seen that the fluid temperature difference is 4 K which is from 296.99 to 300.98 K in the conventional straight fin minichannel. It is observed that the temperature gradient between the near-wall fluid and core fluid is highly developed and the thermal boundary layer keeps increasing as the fluid travels downstream in the conventional straight fin minichannel. This phenomenon deteriorates the convective heat transfer and reduces the cooling effect on the copper surface. However, in Fig. 3.3b, the temperature contour inside the oblique fin minichannel exhibits a more uniform fluid temperature distribution from 298 to 300 K. It is found that a portion of the main flow is diverted into the secondary channel due to the presence of the oblique cuts on the solid fins. This secondary flow, which carries momentum driven by the pressure difference, injects into the adjacent main channel and disrupts the boundary layer as well as accelerates the heat transfer into the core fluid. This results in a better fluid mixing and superior heat transfer performance which lead to lower surface temperature.

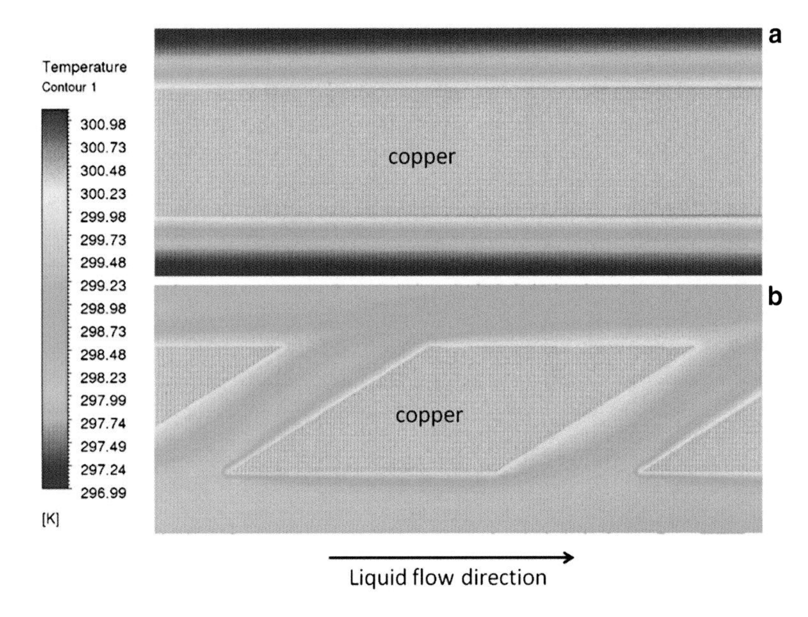

Fig. 3.3 Temperature contour for flow inside (**a**) conventional straight fin minichannel, (**b**) oblique fin minichannel [1]

3.1.2.2 Secondary Flow Distribution

An important phenomenon that affects the heat transfer significantly is how the fluid mixes inside the minichannel. This is a complex physical process which follows the convective diffusion equation which in turn contains the fluid motion terms governed by the Navier–Stokes equations. It is useful to bring the flow field mechanism to account for the heat transfer performance in the cylindrical oblique fin heat sink. Due to experimental limitations, the effects of the secondary flow on the minichannel were studied based on fluid mixing and numerical simulation results. This is feasible since the 3D conjugate simulation predictions generally agree with the experimental results in planar oblique fin heat sink in ref. [2]. Since the oblique fin configuration is periodic, simulation studies focus on flow within a single channel domain instead of the full domain. Similar configuration is adopted for the louvred fin array by DeJong and Jacobi [4].

Figure 3.4 shows the typical cross stream (X, Z) velocity vector and streamline at the middle location in the downstream (X) direction. As fluid is forced into the oblique fin cylinder, coolant travels along the main channel as well as the secondary channel of the cylinder. When the Reynolds number is as low as 50, the streamlines in the secondary channel are uniform and orderly. The velocity is much lower in the secondary channel compared with the main channel. This implies that oblique fin has little effect in flow mixing at the low Reynolds number region.

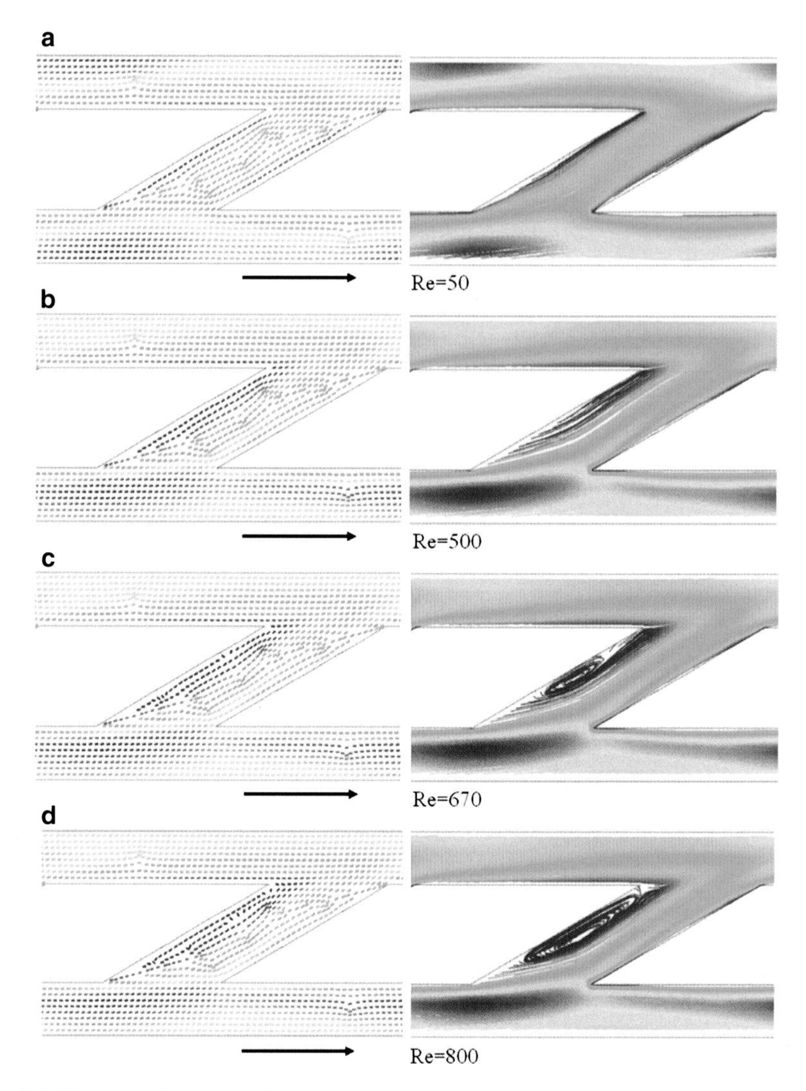

Fig. 3.4 Representative results for the cross stream (*X*, *Z*) velocity vector and streamline at the middle location in the downstream (*X*) direction [1]

When the Reynolds number increases to 500, the streamlines near the trailing edge become rarefaction, but the velocity is still in a relatively orderly pattern. This is due to the secondary flow carrying higher energy and momentum that improves the flow mixing. The flow distribution is non-uniform since there is a slight adverse pressure gradient near the trailing region of the oblique fin. The main channel boundary layer keeps redeveloping at each oblique angle and this enhances the heat transfer performance.

When the Reynolds number increases to 670, the adverse pressure gradient at the trailing edge of the secondary channel enlarges and a recirculation zone whirling in a clockwise direction is formed. DeJong and Jacobi [4] also reported a similar recirculation zone in their studies for louvred fin arrays. This recirculation results in a very high shear stress near the trailing edge of the secondary flow, and this incurs an additional pressure drop since the flow in the recirculation region has high energy that cannot be dissipated.

When the Reynolds number is 840, the flow recirculation is further intensified and is shown as a larger recirculation zone area in Fig. 3.4d. The presence of this region with the higher velocity gradients causes an increment in turbulence and shear stress. The net mass flow region through the secondary channel reduces substantially and this may produce unfavourable effect in flow mixing. As a result, it hinders heat transfer and it suffers a high pressure drop penalty.

3.1.2.3 Entrance Region Effect

The discussion of the fluid behaviour in oblique fin minichannel shows that the velocity profile is disrupted at each entrance of the next downstream fin, and this phenomenon causes the boundary layer growth to be unconstrained from the entrance region. To understand the entrance region effects of the oblique fin minichannel, simulation results based on the velocity streamline and local heat transfer coefficient are employed to understand the hydrodynamic entrance length L_h and thermal entrance length L_t in oblique fin minichannel, respectively, when mass flow rate is 400 ml/min and total heat input is 140 W.

Figure 3.5 shows the velocity streamline profile comparison between conventional straight fin minichannel and cylindrical oblique fin minichannel. For straight fin minichannel, the velocity profile begins to develop along its length until it becomes the fully developed Hagen–Poiseuille velocity profile. The hydrodynamic entrance length L_h is calculated as 0.023 m based on the well-accepted equation [5]:

$$\frac{L_h}{D_h} = 0.05\,\mathrm{Re} \tag{3.1}$$

For oblique fin minichannel, it is observed that the velocity distribution had reached steady state from the second oblique fin. Thus, the pressure drop between each oblique fin unit is calculated and it approaches to a constant value of 3 Pa from 0.008 m. This implies that the friction factors are constant and the velocity profile between each oblique fin unit is steady developed from 0.008 m, which is considerably shorter than the conventional straight fin. This indicates the stable flow behaviour is achieved sooner in oblique fin minichannel rather than straight fin minichannel.

Figure 3.6 presents the local Nusselt number comparison, Nu_x, between conventional straight fin minichannel and cylindrical oblique fin minichannel. It is shown that

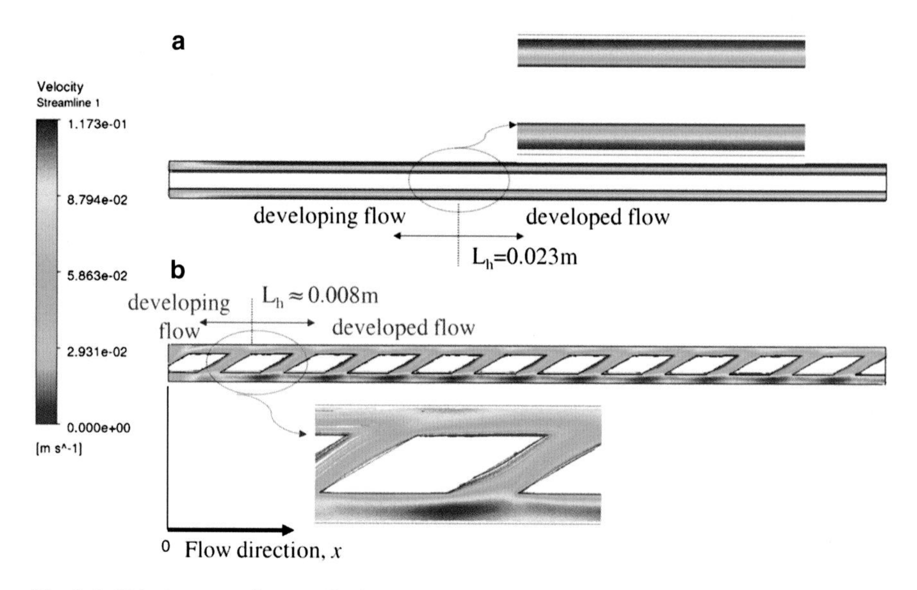

Fig. 3.5 Velocity streamline profile for (**a**) conventional straight fin minichannel, (**b**) oblique fin minichannel

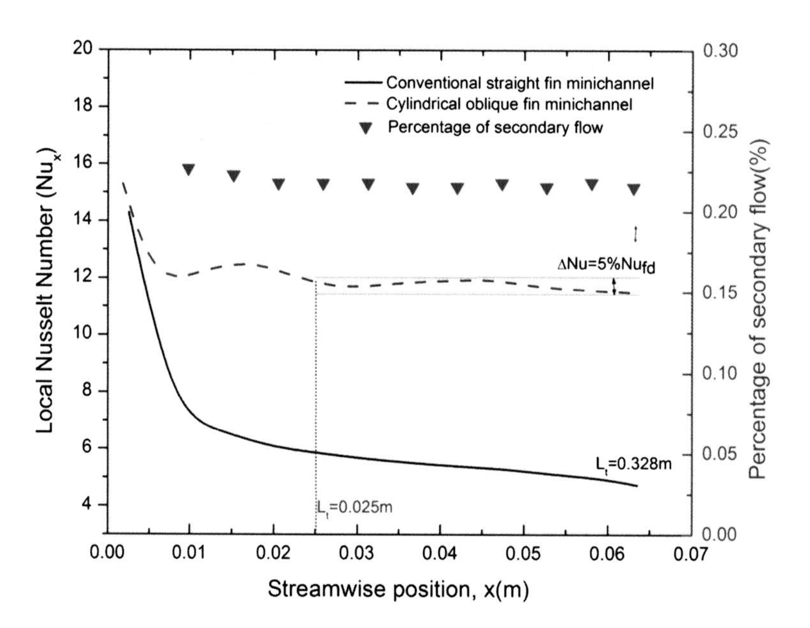

Fig. 3.6 Local Nusselt number comparison between conventional straight fin minichannel and cylindrical oblique fin minichannel

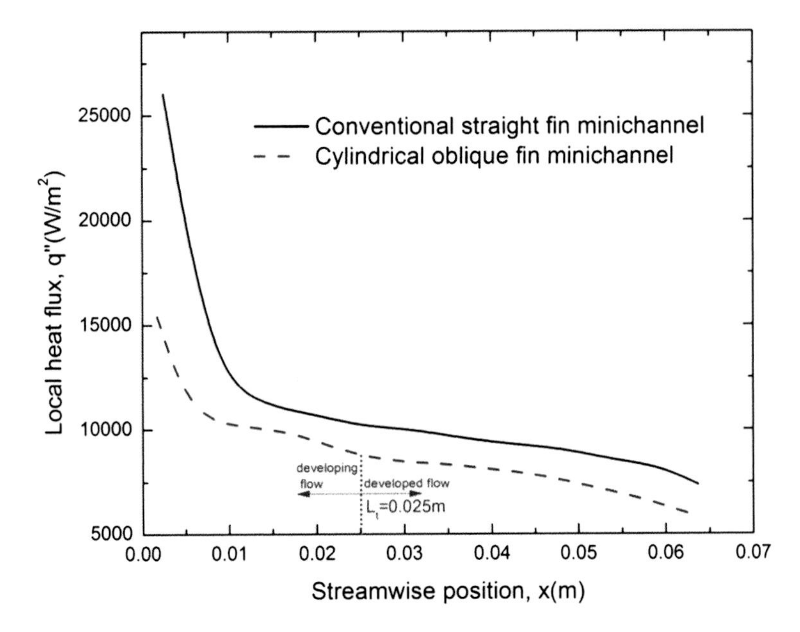

Fig. 3.7 Local heat flux profile between conventional straight fin minichannel and cylindrical oblique fin minichannel

Nu_x for straight fin minichannel changes with streamwise location all the time. Thermal entrance length L_t is calculated as 0.328 m based on the following equation [5]:

$$\frac{L_t}{D_h} = 0.1 \mathrm{Re}\,\mathrm{Pr} \tag{3.2}$$

This provides clear evidence that the conventional straight fin undergoes a thermally developing flow. In contrast, Nu_x for oblique fin minichannel approaches to an average value of 11.8, which can be considered as the steady developed value of Nusselt number, Nu_{fd}. The higher Nu_x can be attributed to the uniform secondary flow generation in cylindrical oblique fin minichannel, which is about 20 % of coolant mass flow across inlet and shown as blue dots in Fig. 3.6. Based on the suggestions by Shah [6], the thermal entrance length L_t is defined as the distance required for the deviation of Nu_x within 5 % of Nu_{fd}. Thus, L_t of oblique fin minichannel in Fig. 3.6 can be determined as 0.025 m, which is significantly shorter than conventional straight fin minichannel. It implies that a thinner thermal boundary layer and steady developed heat flow are achieved sooner due to the shorter entrance length in the cylindrical oblique fin minichannel rather than the straight fin minichannel.

In addition, Fig. 3.7 shows the local heat flux profile between the conventional straight fin minichannel and the cylindrical oblique fin minichannel. It is found that the general trend of heat flux for the cylindrical oblique fin minichannel is lower than the conventional straight fin minichannel due to the larger heat transfer area of

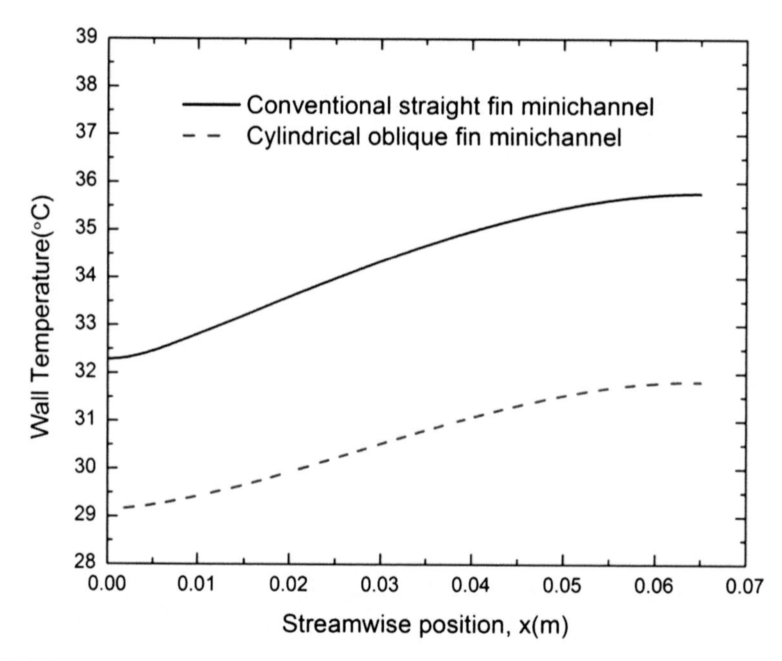

Fig. 3.8 Wall temperature comparison between conventional straight fin and cylindrical oblique fin minichannel

the oblique fin. Conventional straight fin minichannel experiences severe decreasing heat dissipation as it travels downstream. 37 % of the total heat is dissipated at the first 20 % distance travelled in the thermally developing region. Meanwhile, the local heat flux decreasing gradient is more moderate for cylindrical oblique fin minichannel. 48 % of the total heat is dissipated in the developing flow compared with 52 % in developed flow. These results further confirm that cylindrical oblique fin minichannel has more uniform convective heat transfer performance due to the shorter entrance region effect than conventional straight fin minichannel.

3.1.3 *Heat Transfer and Pressure Drop Characteristics*

Due to the secondary flow generation and entrance region effect in cylindrical oblique fin minichannel, superior convective heat transfer performance is achieved, and this results in a more uniform and lower surface wall temperature compared to conventional straight fin minichannel. Figure 3.8 shows the surface wall temperature comparison between conventional straight fin and cylindrical oblique fin minichannel. It is observed that the maximum wall temperature for the oblique fin minichannel is at 31.82 °C and the temperature gradient is at 2.67 °C. In the case of the conventional straight fin minichannel, the maximum wall temperature is at 35.77 °C and the temperature gradient is at 3.48 °C. Therefore, the introduction of oblique fins results in considerable decrease of the maximum wall temperature by 3.95 °C and the temperature gradient by 0.82 °C, respectively.

In addition, Fig. 3.6 shows that the novel cylindrical minichannel induces both notable local and global enhancements. For the conventional straight fin minichannel, the local Nusselt number decreases linearly as the boundary layer thickens when the fluid travels downstream. On the contrary, the local Nusselt number for the cylindrical minichannel with oblique fins nearly reaches a constant value of 11.8 along the downstream direction. This is an enhancement of almost 90 % compared to conventional minichannel. The observation of higher and more uniform heat transfer performance can be attributed to these effects: boundary layer redevelopment, secondary flow effect and shorter entrance length explained in Sect. 3.1.2.

In order to get a better understanding of each surface effect of the oblique fin on flow and heat transfer, the local heat flux and local heat dissipation are calculated based on the simulation results for the fifth oblique fin unit from the inlet location. In Fig. 3.9, the oblique fin surfaces are named as *O1* (oblique fin face 1), *O2* (oblique fin face 2), *M1* (main channel fin face 1) and *M2* (main channel fin face 2), while each bottom wall for the oblique fin unit is named as *BO1* (bottom face 1 of oblique channel), *BO2* (bottom face 2 of oblique channel), *BM1* (bottom face 1 of main channel) and *BM2* (bottom face 2 of main channel). It is seen that *O1* (heat flux 9988.73 W/m^2) and *M1* (heat flux 10574.7 W/m^2) are the most effective heat transfer surfaces among all the faces. They dissipated 56.6 % of the total heat. This is due to the fact that boundary layer is reinitialized and redeveloped, which results in a thinner thermal boundary layer thickness at these two faces. In contrast, the local heat fluxes for faces *O2* and *M2* are 4137.35 and 4255.33 W/m^2, respectively, and they removed 23.07 % of the total heat dissipation. This can be explained by the slightly increased boundary layer thickness on these two faces compared to *O1* and *M1*, which can be seen in Figs. 3.2 and 3.3. Interestingly, the total heat dissipation for the entire bottom surfaces is only at 20.33 %, which is much lower compared to the heat dissipation from the side wall surfaces. Local heat transfer performance shows that *BO1*, *BM1* and *BM2* are the main contributors of heat dissipation for bottom surfaces that removed 19.59 % of the total heat, and the local heat flux is 5,400, 6,665.9 and 3,764.3 W/m^2, respectively. The bottom face *BO2* only dissipated 0.74 % of the total heat and the local heat flux is only 1,179.32 W/m^2. This is because the boundary layer redevelopment at the bottom surface is not as significant as the side wall surface. Thus, lower local heat flux and lower heat dissipation are observed at these bottom surfaces. This is also found in the study of fractal-like branching channel networks by Alharbi et al. [7]. The present study proves that the bottom face *BO2* is the most ineffective heat transfer surface because it is affected by the adverse pressure gradient in the secondary channel significantly.

In addition, one of the most desirable traits for this novel cylindrical oblique fin minichannel is that there is a slight lower pressure drop penalty compared with conventional straight fin minichannel. Figure 3.10 depicts the local pressure profile comparison between conventional and oblique fin minichannel when the mass flow rate is at 400 ml/min. In the conventional channel, it is seen that the continuous

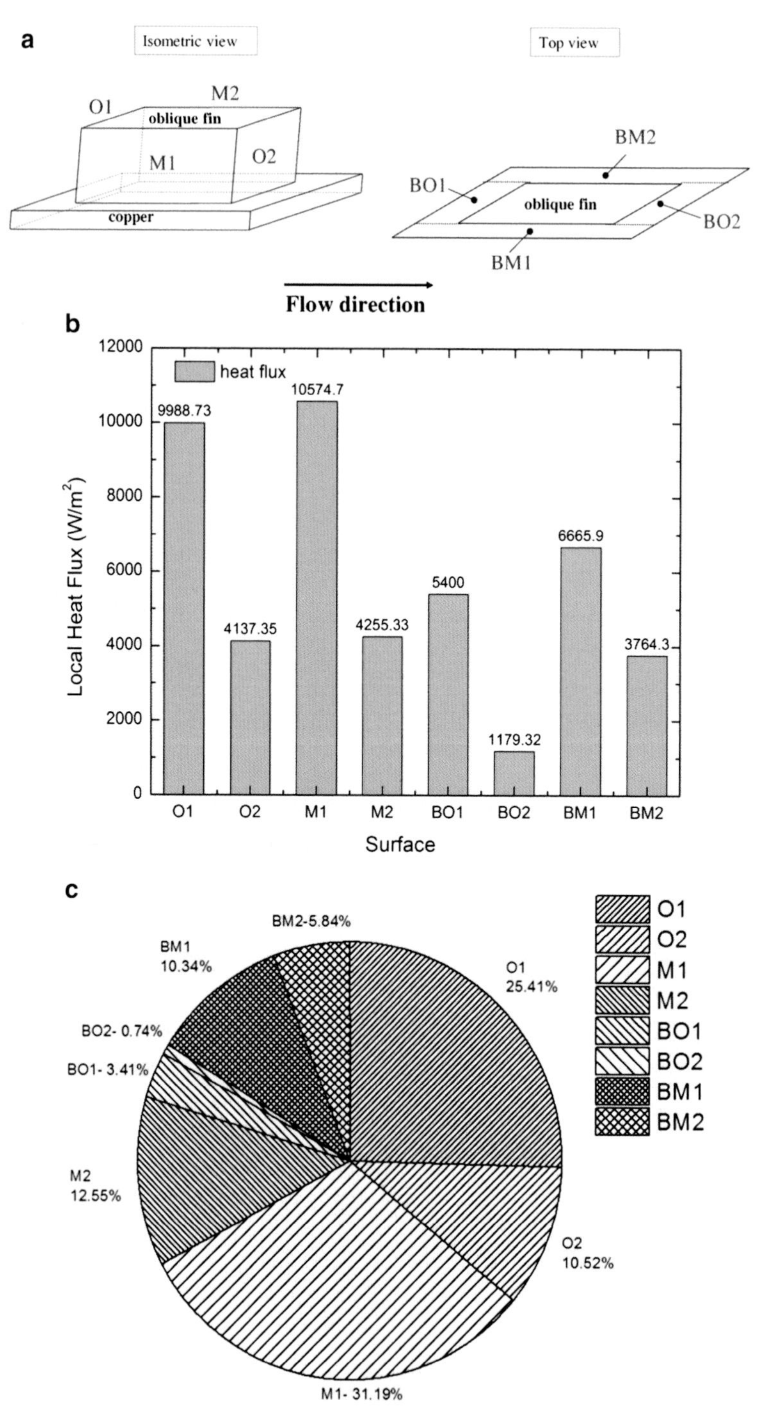

Fig. 3.9 (**a**) Fifth oblique fin unit (not to scale), (**b**) local heat flux on each surface, (**c**) local heat dissipation from each surface

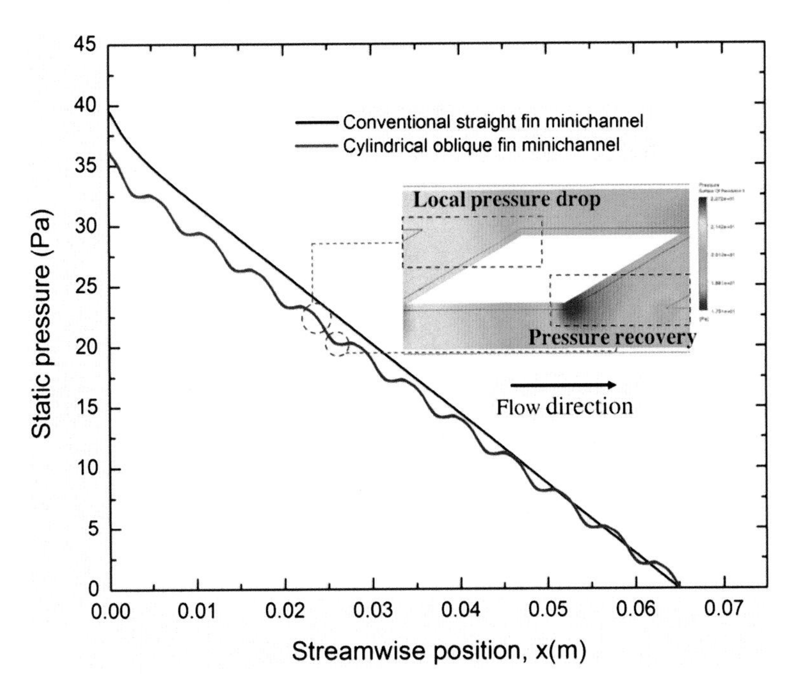

Fig. 3.10 Local pressure profile comparison between conventional straight fin and cylindrical oblique fin minichannel

pressure decreases linearly, and this is due to the friction loss along the channel. On the contrary, the pressure profile of oblique finned microchannel is exemplified as "periodic dips and spikes" and is caused by the different flow behaviours in the oblique minichannel. In Fig. 3.10, it is noticed that the local pressure drops sharply when the secondary flow injects into the main channel at each outlet of the oblique channels. This phenomenon can be attributed to the local friction loss in both oblique channel and main channel; the local reduced cross-sectional flow area also increases local fluid velocity at the expense of pressure. At each entrance of the oblique fins, there is slight pressure recovery to spikes. This can be explained by the local increased cross-sectional flow area in the region of secondary channel gap; thus, the local fluid velocity slows down with increased pressure. This is different from the conventional passive heat transfer enhancement techniques where the heat transfer enhancement is usually accompanied with a large pressure drop penalty.

3.2 Experimental Investigation of Cylindrical Oblique Fin Minichannel Heat Sink

This section provides a description on an experimental investigation of single-phase heat transfer in conventional straight fin minichannel and novel cylindrical oblique fin minichannel. The objective is to validate the applicability of conventional theories and simulation results in predicting heat transfer performance in Sect. 3.1. In Sect. 3.2.1, the test section design, experimental set-up, experimental procedure and data analysis will be presented in detail. Sections 3.2.2–3.2.4 are dedicated to discuss and analyse the heat transfer, pressure drop and overall heat transfer characteristics for both configurations [8].

3.2.1 Experimental Set-Up and Procedures

Figure 3.11 shows the schematic of the flow loop that was configured to circulate cooling liquid through the minichannel test section at desired operating conditions for the experimental set-up. The set-up consists of a reservoir, liquid-to-air heat exchanger, valves, filter, gear pump, flow meter, data acquisition system, DC power supply, computer, pressure transducer, thermocouples and test section where it holds the test piece assembly. A rectangular container of size 12 cm × 15 cm × 20 cm made of polycarbonate material is used for the storage of the deionized water. During the test, deionized water is pumped into the test section through the flow loop using a gear pump (Cole Parmer Console drive 75211–15 with a mount suction shoe pump head 73004–02), and the flow rate is measured using a turbine liquid flow meter (McMillan 104 6T) with a measurement range of 100–1,000 ml/min. Upon exiting the gear pump, a portion of the flow, controlled by a bypass valve, entered the flow loop, while the remaining portion returned to the reservoir through a bypass loop. A needle valve is used to control and adjust the flow for the loop. Temperature measurements are obtained at the inlet and outlet plenum of the test

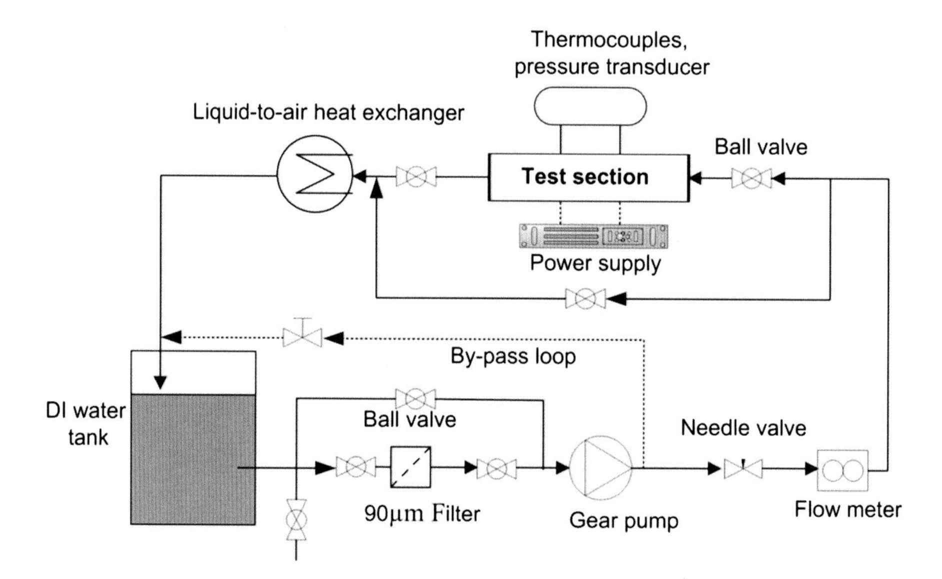

Fig. 3.11 A schematic flow loop of experimental set-up

section as well as at another eight locations below the channel surface of the test piece using T-type Watlow thermocouples, respectively. The pressure drop between the inlet and outlet plenum of the test section is measured using GE LP 1000 Series Druck Ultra Low differential pressure transmitter (model, LPM1512C1SNW1). The test section is heated using a cartridge heater which is powered by a 850 W programmable DC power supply (Sorensen XTR150-5.6) with an output range of 0–150 V and 0–5.6 A. In order to maintain a constant temperature of the water in the reservoir after it is pumped out from the test section, a Thermatron liquid-to-air heat exchanger (model, 735SPC2A01) is used to regulate the desired temperature before the water is pumped back into the reservoir. Swagelok stainless steel pipes and fittings were used to construct the flow loop. The data from all different sensors were collected using a National Instruments high-speed data acquisition system (16-channel thermocouple input module NI 9214 and 16-channel ±20 mA/±10 V input module NI 9207).

The test section consists of four parts, namely, the housing, the cover, the top adaptor and the copper block microchannel heat sink shown in Fig. 3.12. The housing comprises of the top housing, the bottom housing and the main housing, all of which are made of Teflon. The top housing holds the top adaptor, top cover and microchannel heat sink. It has two O ring slots, one within the top plate and the other at the top housing to prevent leakage. At the top and bottom housing, there are

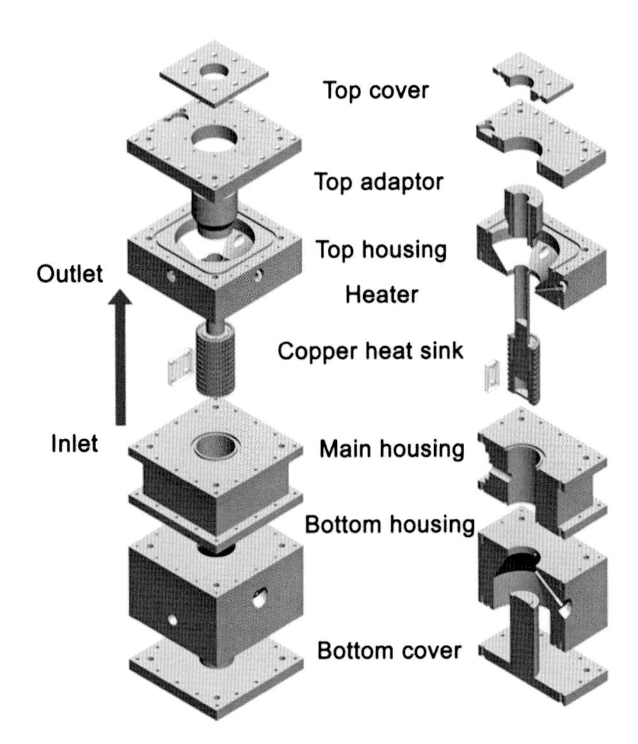

Fig. 3.12 Exploded view of test section

independent pressure and temperature ports for measuring the fluid properties before and after bypassing the heat sink. The minichannel heat sink is made from a copper block in which minichannels are cut on the surface using computer numerical control (CNC) machining process. There are eight holes adjacent to each other around the circumference below the channel surface in the block for inserting the thermocouples to measure the heat sink's streamwise temperature distribution. These eight holes were drilled 4.5 mm below the channel surface: 13, 26, 39 and 52 mm below the outlet plenum, respectively. The bottom housing was used to hold the bottom cover and minichannel heat sink and provided uniform flow to the inlet channels.

Experiments were carried out on the heat sinks with conventional straight fin minichannel and novel cylindrical oblique fin minichannel. The actual test pieces are shown in Fig. 3.13. The detailed dimensions for both are given in Table 3.2.

Fig. 3.13 Experimental test pieces of conventional straight fin and cylindrical oblique fin heat sinks [8]

Table 3.2 Geometrical details for conventional straight fin and cylindrical oblique fin minichannels [8]

Characteristic	Conventional straight fin minichannel	Oblique fin minichannel
Material	Copper	
Footprint dimension D, (mm)	18	
Fin length L, (mm)	65	
Number of rows	36	
Channel height H, (mm)	2.002	2.039
Main channel bottom width (mm)	1.365	1.532
Main channel top width (mm)	1.730	1.973
Fin width (mm)	1.598	1.542
Aspect ratio, α	1.47	1.31
Number of fins per row	–	11
Secondary channel length (mm)	–	2.008
Oblique fin length (mm)	–	3.744
Oblique angle θ (°)	–	30.8

In order to obtain accurate data of the temperature, pressure drop and velocity for the flow through the test sections, the thermocouples, pressure sensor and flow meter were carefully calibrated before the commencement of each experiment. The experimental procedure, data reduction and uncertainty analysis can be found in Sect. 2.3. The present experiments were conducted at flow rate ranging from 50 to 900 ml/min and heat input is varied from 50 to 300 W.

3.2.2 *Heat Transfer Characteristics*

The experimental investigation on both conventional and cylindrical oblique fin minichannel heat sinks is conducted over the flow rates that ranged from 50 to 900 ml/min, which correspond to Reynolds numbers of 50–500 and with the heat input that ranged from 50 to 300 W. Since $L/\mathrm{Re}D_h > 0.05$ (hydrodynamically fully developed) and $L/\mathrm{Re}D_h Pr < 0.05$ (thermally developing), all the experimental data correspond to the thermally developing regime criterion for conventional minichannel.

The measured performance is presented and compared with the numerical predictions. Figure 3.14 shows the wall temperature distribution on the heat sink surface at different streamwise locations when the flow rate is at 400 ml/min and heat input is 170 W. The continuous lines are obtained from simulation prediction in Sect. 3.1, while the dots are obtained from experiments. It is found that the deviation between experimental and numerical results is less than 6 % under all conditions. This means the numerical simulation studies in Sect. 3.1 are validated by the present experiments. It is also observed that the wall temperature in the heat sink is increased along the streamwise location. This is due to the increasing fluid bulk mean temperature which could determine the energy balance when the fluid travels in the downstream.

A greater local heat removal capability can be achieved by using the cylindrical oblique fin heat sink in comparison with the straight fin heat sink. Figure 3.15 shows the local Nusselt number at different locations for both conventional straight fin and cylindrical oblique fin heat sink when the Reynolds number is 310. It is observed that cylindrical oblique fin heat sink has a consistently higher Nusselt number compared to conventional heat sink at all locations within the heat sink. At the first location which is 13 mm from the flow entrance, the local Nusselt number enhancement is as high as 126 % compared with the conventional heat sink. The minimal local Nusselt number enhancement is 57 %. The mechanism behind this heat transfer enhancement is that the thermal boundary layer is constantly being reinitialized at leading edge of each oblique fin which causes the fluid flow to remain in the developing state and the thermal boundary layer becomes thinner for the oblique fin minichannel. On the other hand, the thermal boundary layer thickness keeps increasing as it travels downstream until it becomes fully developed for the case of the straight fin minichannel.

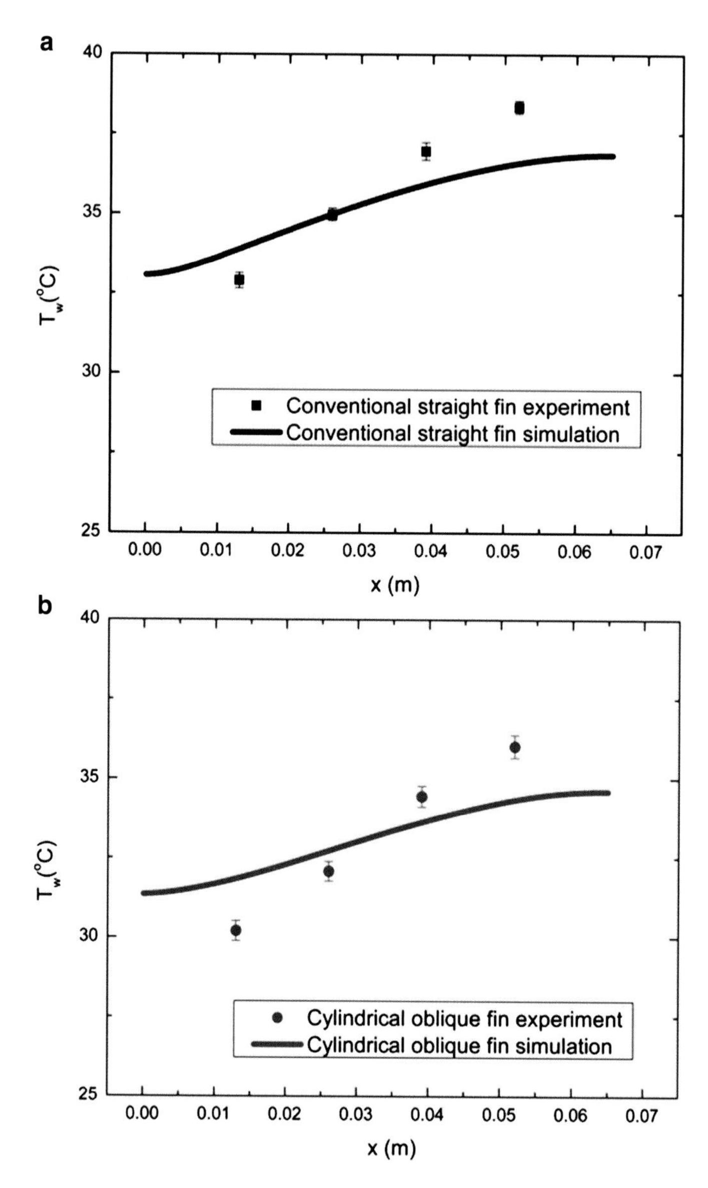

Fig. 3.14 Wall temperature comparison of (**a**) conventional straight fin minichannel, (**b**) cylindrical oblique fin minichannel

Figure 3.16 shows the graph of the average experimental and numerical Nusselt number against Reynolds number for both conventional and cylindrical oblique fin minichannel heat sinks. The experimental values of the Nusselt number are derived from the average values from the eight thermocouples at the stated flow rate. Both experimental data obtained in the two test modules show that the trend of the water

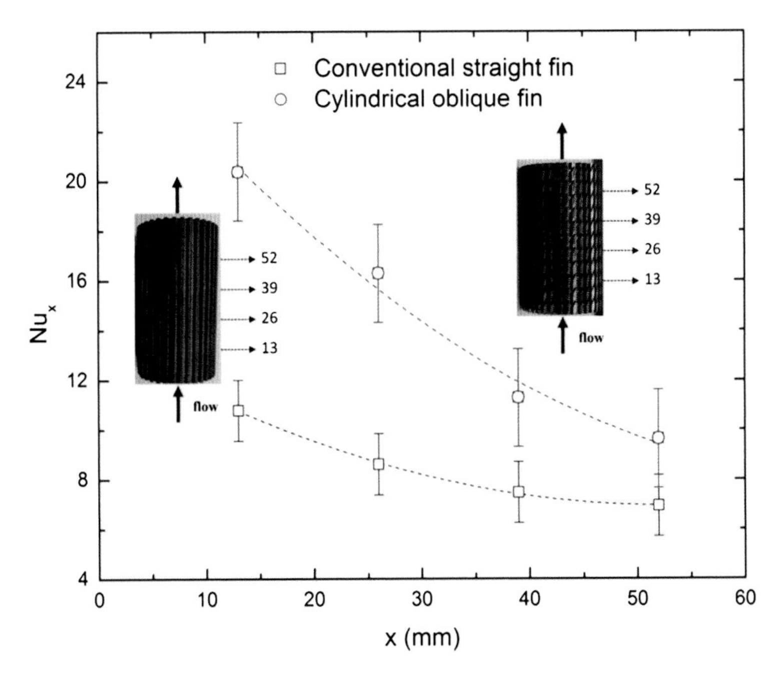

Fig. 3.15 Local Nusselt number for conventional straight fin and cylindrical oblique finned minichannel heat sink

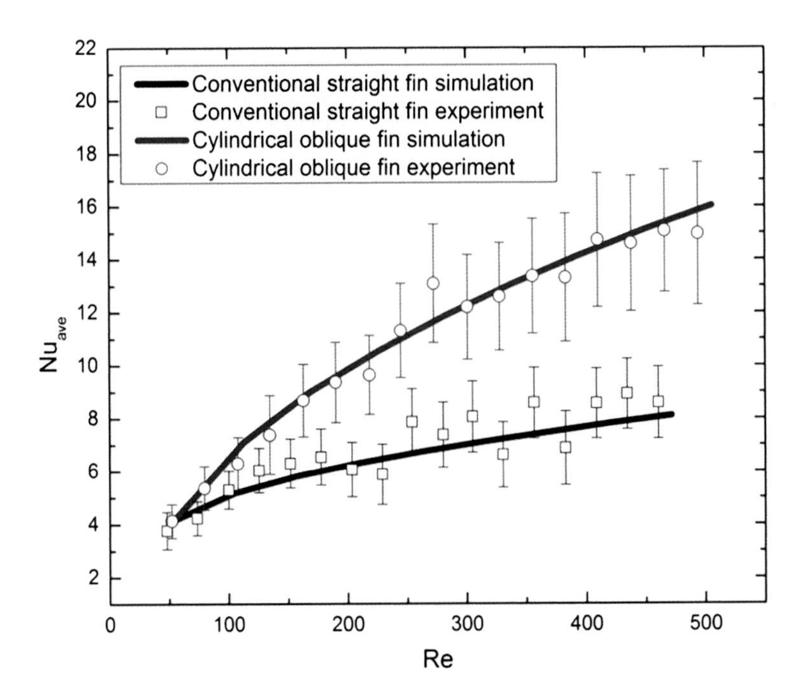

Fig. 3.16 Average Nusselt number obtained from experiments and numerical analyses for conventional straight fin and cylindrical oblique finned minichannel heat sink

flow in minichannel is similar with that of the prediction of the simulation results. It can be seen that the average Nusselt number for both minichannel heat sinks increases with Reynolds number because the thermal boundary layer thickness decreases with increased fluid velocity. Nevertheless, the heat transfer performance for the minichannel with cylindrical oblique fin is significantly higher than conventional minichannel heat sink. The average Nusselt numbers for both configurations are almost equivalent at the lower Reynolds number of 50 since the flow is considered zero and convection heat transfer is negligible. However, the average Nusselt number increases to as much as 75.64 %, from 8.58 to 15.07, when the Reynolds number reaches 460. This heat transfer performance increment is 20 % higher than that of planar oblique fin microchannel [2]. This noticeable enhancement in heat transfer is due to the combined effects of thermal boundary layer redevelopment at the leading edge of each cylindrical oblique fin and the uniform secondary flows generated by flow diversion through the oblique fins as explained in Sect. 3.1.

The total thermal resistance comprises of conductive, convective and caloric thermal resistance. The conductive thermal resistance is greatly dependent on the heat sink material property, and both use the same copper material with a thermal conductivity of 387.6 W/m K. Thus, the conductive thermal resistance is the same for both heat sinks. The caloric thermal resistance reduces with increasing flow rate; however, it is not a significant term in liquid cooling system since ρc_p is very high and has little effect on the thermal resistance. The convective thermal resistance reduces with increasing Reynolds number and results in lower total thermal resistance.

Figure 3.17 shows the graph of the experimental and numerical total thermal resistance against Reynolds number for both conventional heat sink and cylindrical oblique fin heat sink. As shown in Fig. 3.17, the experimental and numerical result matches closely and the differences are all within 12 % tolerance. As the flow rate increases, the Reynolds number rises and the total thermal resistance in minichannel decreases exponentially. This is because the thermal boundary layer thickness decreases as the fluid velocity increases. It is found that the total thermal resistance of the cylindrical oblique fin minichannel ($R_{tot}=0.029$ °C/W) reduced by as much as 59.1 % compared with conventional minichannel ($R_{tot}=0.046$ °C/W) when the Reynolds number was around 460.

In the experiment, the lower total thermal resistance contributes to the lower surface temperature of the minichannel heat sink. Similarly, a comparison of wall temperature at the same location between cylindrical oblique fin and conventional heat sinks was made at a heat flux of 6.1 W/cm^2 and a Reynolds number of 310. A reasonable constant heat transfer performance was observed across the cylindrical oblique fin minichannel heat sink. As shown in Fig. 3.18, for all the test points, the cylindrical oblique fin heat sink has a lower wall temperature compared with the conventional heat sink. A nearly uniform temperature drop of 4.3 °C is observed in the streamwise direction.

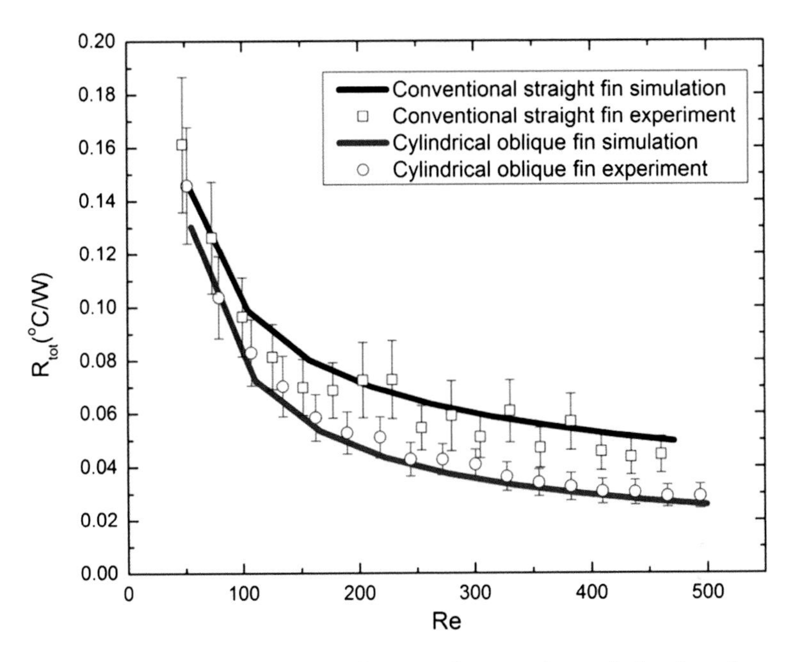

Fig. 3.17 Total thermal resistance obtained from experiments and numerical analyses for conventional straight fin and cylindrical oblique finned minichannel heat sink

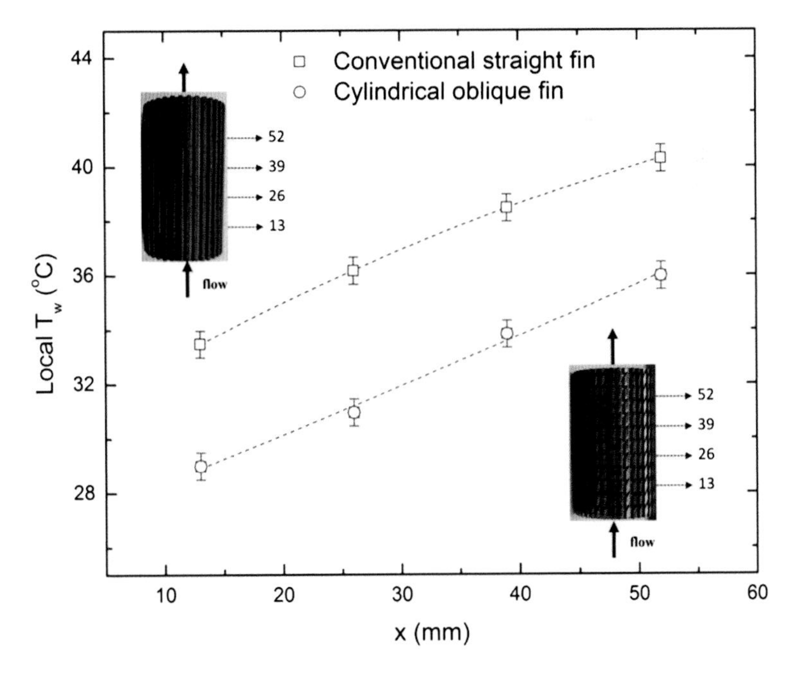

Fig. 3.18 Local wall temperature distribution for conventional straight fin and cylindrical oblique fin minichannel heat sink

3.2.3 Pressure Drop Characteristics

Figure 3.19 plots the pressure drop comparison between inlet and outlet of the two heat sinks. The solid line shows the simulation results, while all the discrete data represent the experimental data for the thermally developing laminar flow at Reynolds number from 50 to 500 for both conventional and cylindrical oblique fin minichannel heat sinks. The experimental data obtained in the test modules shows that the pressure drop trend in the two minichannels is similar with that of the prediction of the simulation results. The pressure drop deviation is higher than expected for both minichannel heat sinks at lower Reynolds number region. This discrepancy was thought to be due to the uncertainties in channel dimensions, surface roughness and flow rate measurement errors which is reported in ref. [9]. The pressure drop increases as the Reynolds number increases because as the coolant flow rate increases, the more frictional loss it faces as it transverses through the heat sink. However, a significant heat transfer augmentation of the cylindrical oblique fin minichannel heat sink is achieved with a small pressure drop penalty compared with the conventional minichannel. At low Reynolds number, the pressure drop for conventional heat sink is a little bit higher than the cylindrical oblique fin configuration. This is probably because most of the coolant flows through the main channel with very little flow going through the secondary channels. As the Reynolds number increases, a higher percentage of coolant is diverted into the oblique channels.

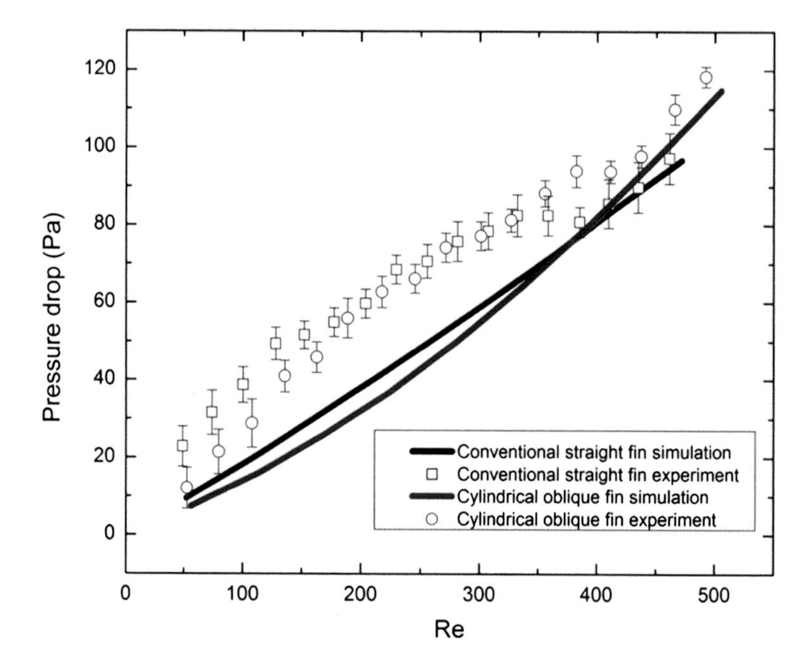

Fig. 3.19 Pressure drop for conventional straight fin and cylindrical oblique finned minichannel heat sink

This creates a stronger secondary flow which further augments the heat transfer but incurs additional pressure drop penalty. Therefore, the pressure drop for oblique fin minichannel starts to deviate and increases more than the conventional configuration when Reynolds number is beyond 350. As shown in Fig. 3.19, when the Reynolds number is around 460, the maximum pressure drop of the cylindrical oblique fin minichannel is 106 Pa and for the conventional minichannel it is 97 Pa. Therefore, the cylindrical oblique fin heat sink generates secondary flow which enhances its heat transfer performance yet maintains a comparable pressure drop compared with the conventional heat sink.

3.2.4 Overall Heat Transfer Characteristics

Figure 3.20 shows the average heat transfer enhancement and pressure drop penalty at different Reynolds number for the conventional straight fin channel and cylindrical oblique fin channel. The heat transfer enhancement (E_{Nu}) and pressure drop penalty (E_f) are defined as the average Nusselt number and friction factor of the cylindrical oblique fin channel divided by that of conventional straight fin channel, respectively [10]. As shown by the E_{Nu} line, the value is always higher than *1* which implies that the oblique fin channel is superior to conventional straight fin channel in heat

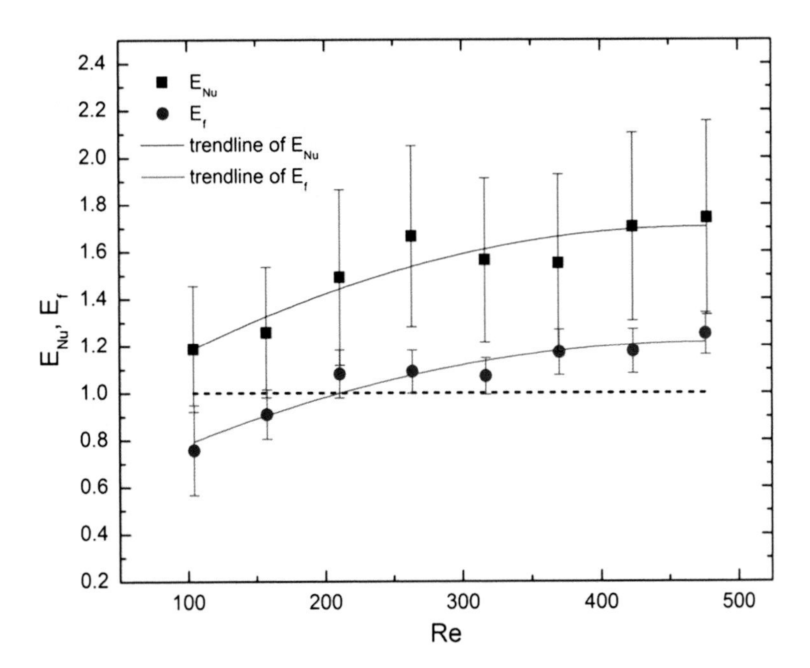

Fig. 3.20 Average heat transfer enhancement and pressure drop penalty for different Reynolds number

transfer performance. For the case of the line E_f, at Reynolds number from 50 to 200, the overall friction factor of the oblique fin minichannel is lower than the conventional straight minichannel; however, at higher Reynolds number, the friction factor for the conventional minichannel is lower. It should be noted that at higher Reynolds number, the heat transfer performance is improved by about 74 % for the oblique fin minichannel over the conventional straight minichannel, while the friction factor increases only about 20 %. This shows that the cylindrical oblique fin minichannel can improve energy efficiency significantly and save more pumping power overall.

3.3 Edge Effect Investigation of Cylindrical Oblique Fin Heat Sinks

It was shown in previous sections that cylindrical minichannel heat sink with oblique fins is an effective cooling technique with enhanced flow mixing. However, it is desirable to understand how the edge effect influences the flow and temperature uniformity due to the generation of secondary flows in the minichannels. This section examines the edge effect for oblique fin structure on both "planar" and cylindrical heat sinks. The introduction of edge effect is firstly presented in Sect. 3.3.1. The procedures to establish the experimental set-up and numerical simulation approaches are illustrated in Sect. 3.3.2. Flow and temperature distributions are quantified along both spanwise and streamwise directions of the heat sinks through numerical studies, while edge effect and temperate uniformity characteristics are investigated in Sect. 3.3.3. These results provide insights for designers interested in oblique finned heat sink or other similar designs. Furthermore, the results provide guidelines for researchers attempting to develop and simulate full-scale model experiments.

The phenomenon being investigated in this section is the edge effects presented in the planar oblique fin microchannel design. In Fig. 3.21, it is shown that due to the non-symmetrical nature of the oblique channels, one side of the heat sink experiences draining of coolant along the edge, while the opposite edge has more coolant flowing in and filling it. The edge that has draining tendency is now labelled as draining edge, while the opposite edge that allows more water entering is the filling edge. There is a net flow migration across the planar heat sink footprint. This is denoted as edge effect in this study for clarity and it has a significant effect on heat transfer. Besides, edge effects due to flow maldistribution may induce non-uniform temperature distributions along the heat sink footprint (quantitative analysis in Sect. 3.4). If it was used to cool an electrical component, the non-uniform temperature distribution might cause uneven thermal expansion of the device and could damage its electrical properties [11]. A similar phenomenon of bounding wall effects on flow and heat transfer in louvred fin arrays was investigated by DeJong and Jacobi [4]. They found that the flow characteristics near walls have competing effects on

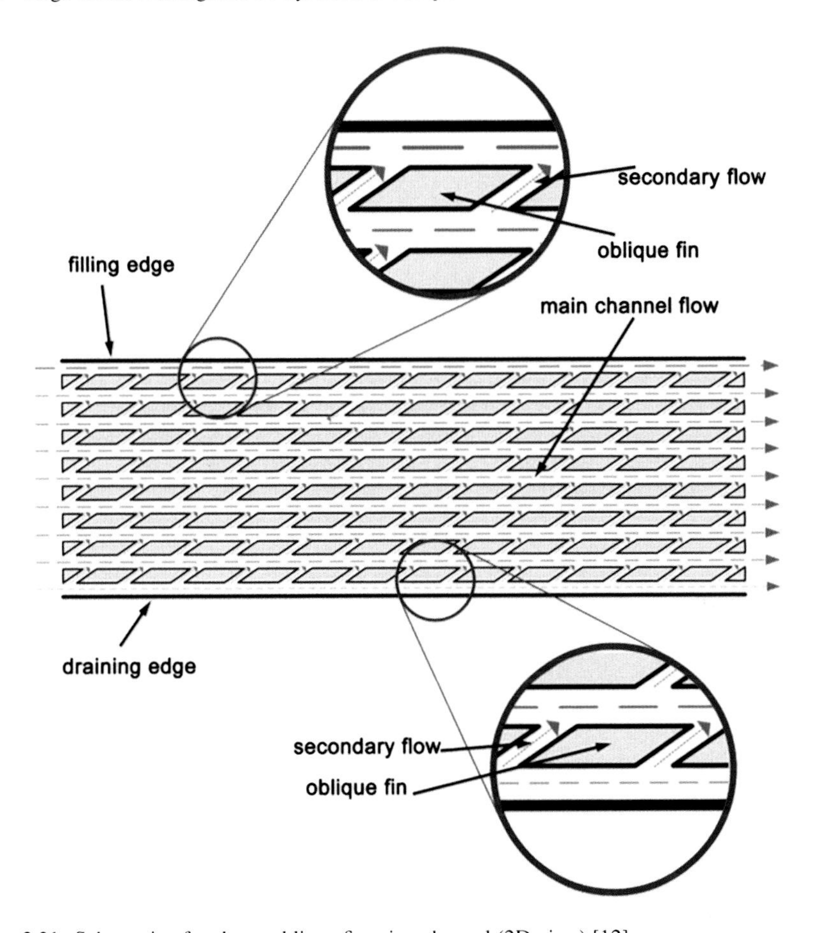

Fig. 3.21 Schematic of a planar oblique fin microchannel (2D view) [12]

heat transfer and have a profound effect on fluid flow in louvred fin arrays. Lee et al. [2] employed sectional oblique fins to improve the flow mixing and enhance heat transfer performance on planar heat sink surface. It was reported that there was a slight deviation between numerical simulation and experimental results and it is attributed to the different flow behaviours between the side edge and the middle regions of the heat sink.

Therefore, this study is to scrutinize the edge effect on flow and temperature uniformity in minichannel heat sink. Experimental studies were performed to quantify the temperature uniformity characteristic and local temperature distribution on both "planar" and cylindrical oblique fin heat sinks. Moreover, numerical simulation models based on the real test pieces were adopted to provide an improved understanding of the influence of edge effects on flow field and temperature distribution uniformity.

3.3.1 *Experimental Procedures and Numerical Approach*

3.3.1.1 Experimental Set-Up and Procedures

Details of experimental set-up, test section, test procedure, heat loss calculation and uncertainty calculation are described in Sect. 3.2. Besides, in order to improve the testing accuracy, a water bath instead of reservoir is used for adjusting and storage of the deionized water with a constant inlet temperature of 21 °C for all the testing in this study [12].

The cylindrical minichannel heat sink was made from copper in which the minichannels with oblique fins were machined using computer numerical control (CNC) machining process. The cylindrical heat sink without "edge effect" is the regular cylindrical oblique fin minichannel heat sink. The cylindrical heat sink with "edge effect" was formed by blockading seven straight channels with an insulation material (modelling clay). Figure 3.22a shows the configuration of blockaded cylindrical oblique fin heat sink. The blockade of the channel was to induce the "edge effect" by not allowing fluid to interact circumferentially. In Fig. 3.22b, seven holes were drilled around the circumference of the heat sink for inserting the thermocouples to measure the heat sink's spanwise temperature distribution. First, these seven holes were drilled 4.5 mm below the channel surface, as 26 mm below the outlet plenum. After measuring the spanwise temperature distribution for both configurations, they were drilled further to 39 mm below the outlet plenum, and measurements were taken again using the T-type thermocouple. The oblique fin angle points out from location 1. This indicates that this is the draining edge while location seven is the filling edge.

Experimental and numerical investigations were performed on both blockaded and un-blockaded cylindrical minichannel heat sinks with oblique fins.

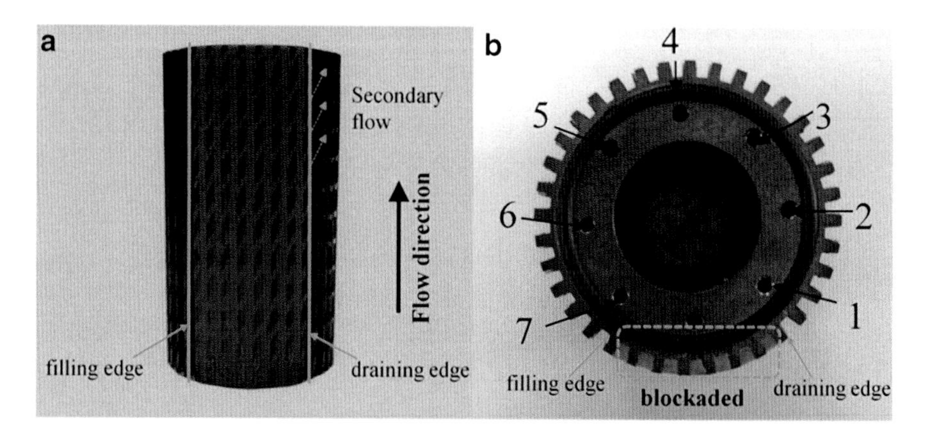

Fig. 3.22 (**a**) Blockaded test piece configuration, (**b**) cross-sectional view and thermocouple locations on blockaded test piece

The detailed geometrical dimensions for both configurations can be found in Table 3.2. Experiments were conducted at flow rates of 100, 500 and 800 ml/min and heat inputs were 100, 200 and 300 W.

3.3.1.2 Numerical Simulation Approach

The commercial CFD software ANSYS CFX is adopted in this study to solve the Navier–Stokes equations using a fully conservative, element-based finite volume method. To ensure robustness, CFX provides flexibility in choosing discretization schemes for each governing equation. The discretized equations, along with the initial condition and boundary conditions, were solved using the coupled method. Using the coupled solver, the hydrodynamic equations are solved as a single system. Pressure–velocity coupling is enforced with a non-stagger grid and the fourth-order accurate algorithm of Rhie and Chow [13].

In the present study, steady Reynolds-averaged Navier–Stokes (RANS) equations for turbulent incompressible fluid flow with constant properties are used. The governing flow field equations are the continuity and the RANS equations, which are given by

$$\frac{\partial u_j}{\partial x_j} = 0 \tag{3.3}$$

$$\frac{\partial u_i u_j}{\partial x_j} = -\frac{1}{\rho}\frac{\partial p}{\partial x_i} + \frac{\partial}{\partial x_j}\left(\nu S_{ij} - \overline{u_i' u_j'}\right), \quad j = 1, 2, 3\ldots \tag{3.4}$$

where S_{ij} is the main strain rate and calculated by $S_{ij} = \frac{1}{2}\left(\frac{\partial u_i}{\partial x_j} + \frac{\partial u_j}{\partial x_i}\right)$ and $\overline{u_i' u_j'} = \tau_{ij}$ is Reynolds stress tensor, while u_i' represents the velocity fluctuation in i-direction. These terms arise from the non-linear convection in the un-averaged equation, and they reflect the fact that convective transport due to turbulent velocity fluctuations will act to enhance mixing over and above that is caused by thermal fluctuations at the molecular level.

The shear stress transport k–w model (SST) is employed to predict the flow behaviour in the present study. It is based on the following two transport equations:

$$\frac{\partial(\rho u_i k)}{\partial x_i} = \gamma P_k - \beta_1 \rho k \omega + \frac{\partial}{\partial x_i}\left[\left(\mu + \frac{\mu_t}{\sigma_k}\right)\frac{\partial k}{\partial x_i}\right] \tag{3.5}$$

$$\frac{\partial(\rho u_i \omega)}{\partial x_i} = \alpha \rho S^2 - \beta_2 \rho \omega^2 + \frac{\partial}{\partial x_i}\left[\left(\mu + \frac{\mu_t}{\sigma_{\omega 1}}\right)\frac{\partial \omega}{\partial x_i}\right] + 2(1 - F_1)\rho \frac{1}{\sigma_{\omega 2}\omega}\frac{\partial k}{\partial x_i}\frac{\partial \omega}{\partial x_i} \tag{3.6}$$

where F_1 is a blending function, and it is designed to be one in the near-wall region which activates the standard k–w model and zero away from the wall which activates the transformed k–ε model. The model also includes a slight amendment to the eddy viscosity for a better prediction of the turbulent shear stress. More details of the SST model can be found in Sparrow et al. [14], Menter [15] and Lee et al. [16]. It could give a highly accurate prediction of the onset and the amount of flow separation under adverse pressure gradients by the inclusion of transport effects into the formulation of the eddy viscosity. The superior performance of this model has been demonstrated in a large number of validation studies [17].

3.3.2 *Flow Field Analysis*

For the planar oblique fin structure, edge effects have profound effect on heat transfer and can cause hotspots near the corners of the heat sink. Both experimental and numerical studies were conducted based on the blockaded test piece (similar structure as planar oblique fin test piece) and un-blockaded test piece (regular cylindrical oblique fin structure). The mass flow rates are set as 100, 500 and 800 ml/min and heat inputs were set as 100, 200 and 300 W.

3.3.2.1 Validation of Numerical Simulations

Figure 3.23 shows the wall temperature comparisons for cylindrical oblique fin minichannel with different simulation models and experimental studies when the flow rate is 400 ml/min (Re = 216) and heat input is 170 W. The continuous solid line is obtained from full domain simulation with SST model using CFX. The continuous dot line is obtained from simplified model simulation with laminar model using Fluent (in Sect. 3.1), while the dots are obtained from experimental measurement for cylindrical oblique fin minichannel testing. It is found that the maximum temperature deviation between experimental and numerical results is less than 6 % under all conditions. The numerical results obtained from SST model for full domain study almost match those obtained from Fluent using laminar model in the simplified study since the maximum temperature deviation is only 1.2 %. This indicates that the numerical simulation studies with both SST model and laminar model are validated by the experiments.

3.3.2.2 Flow Distribution Study

Due to the nature of oblique structure, edge effects exist in the planar oblique fin heat sink. This is demonstrated by blocking seven channels of the cylindrical oblique fin heat sink to form the blockaded cylindrical heat sink. Figure 3.24a shows the full domain velocity profile of the minichannel when the flow rate is 800 ml/min at a total heat input of 300 W. It can be seen that non-uniform flow distribution area is found in the blockaded heat sink. At the inlet region of the heat sink, large amount

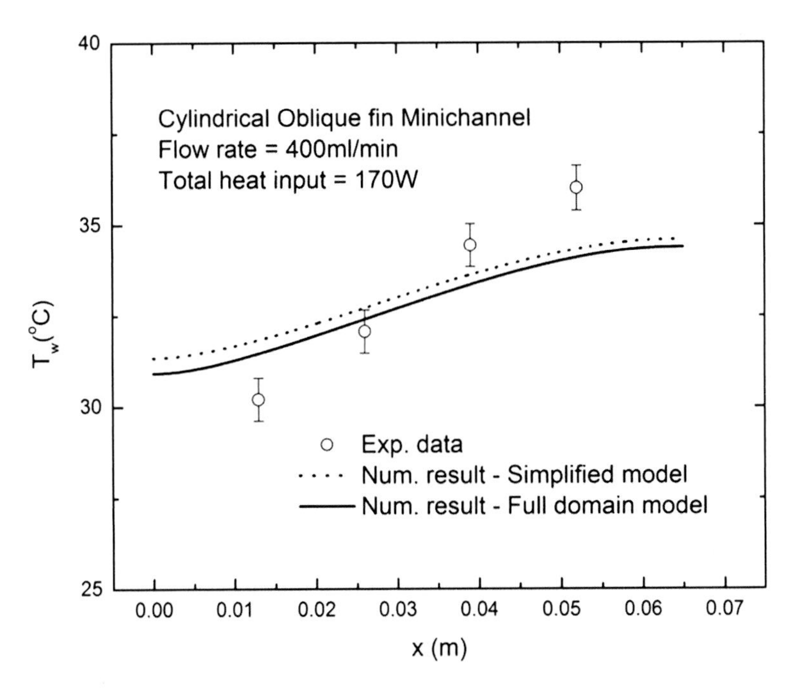

Fig. 3.23 Local wall temperature comparison for cylindrical oblique fin minichannel heat sink

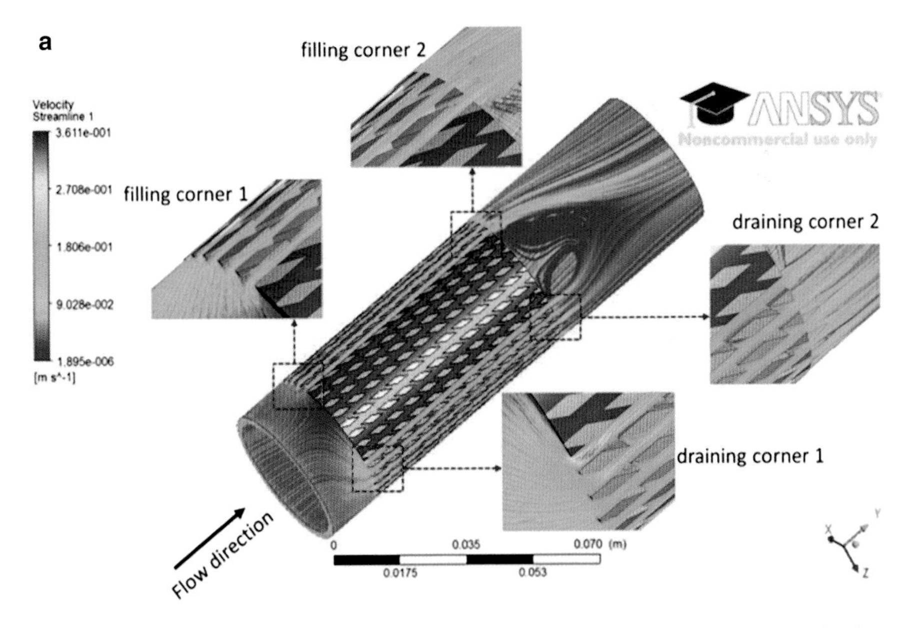

Fig. 3.24 Flow distribution of the blockaded cylindrical heat sink: (**a**) velocity contour for flow field domain, (**b**) main channel mass flow rate for draining edge and filling edge, (**c**) main channel mass flow rate, (**d**) secondary channel mass flow rate [12]

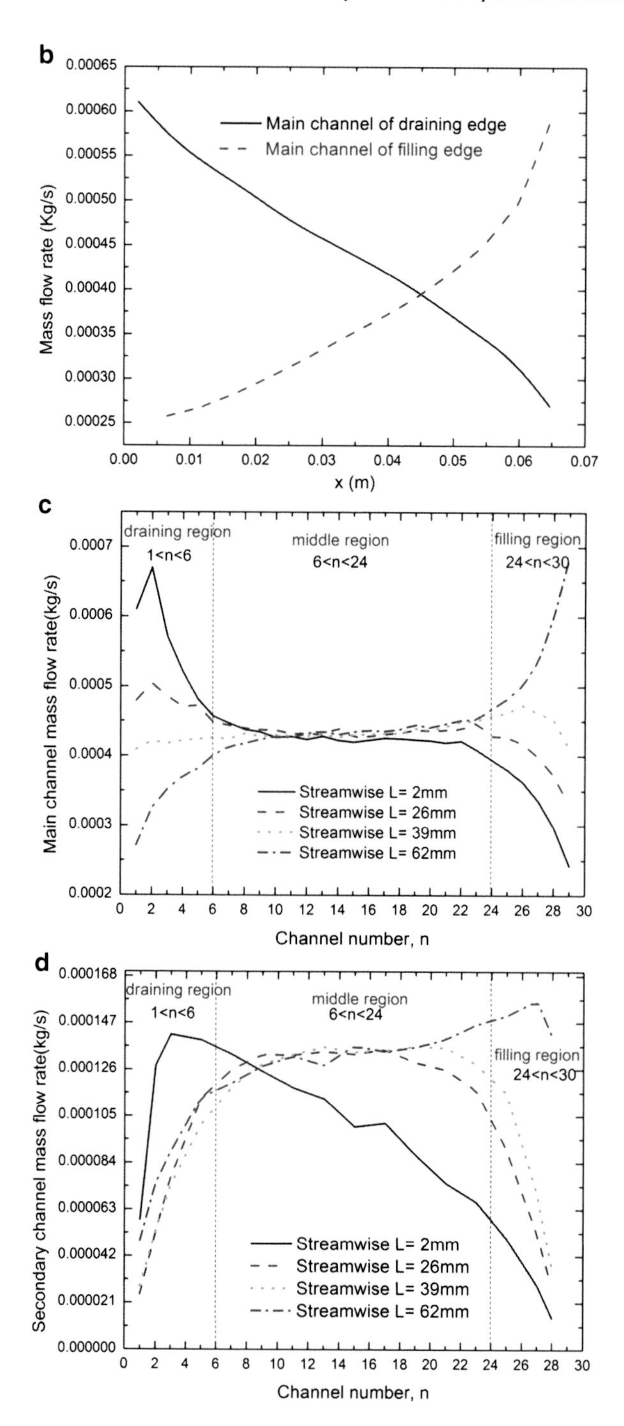

Fig. 3.24 (continued)

of flow is redistributed, and this results in higher mass flow through the draining corner 1 and lower mass flow through the filling corner 1. Figure 3.24b shows that the main channel mass flow rate of draining edge is progressively decreasing from 0.0006 to 0.00027 kg/s along the streamwise location. Conversely, due to the combined effect of oblique fin nature and flow expansion at outlet region, there is more flow accumulated at the filling corner 2 rather than that at filling corner 1. In Fig. 3.24b, the main channel mass flow rate is increasing from 0.00024 to 0.0006 kg/s along the streamwise direction. This results in the non-uniform velocity profile development and compromised heat transfer performance.

In order to quantify the flow distribution inside the blockaded cylindrical heat sink, the mass flow rates in main channel and secondary channel are calculated respectively. Figure 3.24c, d shows the fluid flow distribution in the main and secondary channels for blockaded cylindrical heat sink. From Fig. 3.24c, it can be seen that there are three different flow regions in the full fluid domain study, namely, the draining region ($1<n<6$), middle region ($6<n<24$) and filling region ($24<n<30$), while n is denoted as the channel number starting from draining edge. In the draining region ($1<n<6$), the main channel mass flow rate in the upstream (2 and 26 mm) is decreasing and that in the downstream (39 and 64 mm) it is increasing to a stable main flow rate of 0.00043 kg/s. Even though secondary mass flow rate increases to a stable flow rate of 0.00013 kg/s, the flow mixing is very poor because the maximum secondary mass flow is only 23.2 % of the total channel flow. Thus, hotspots appear in this region and become more severe along the streamwise direction. Interestingly, a stable main channel mass flow and stable secondary mass flow are found in the middle region in both spanwise and streamwise location ($6<n<24$). The main channel flow in spanwise location (upstream) shows a decreasing trend, and this is due to the flow redistribution caused by the inlet flow section area contraction, while the secondary channel flow is increasing slightly in the streamwise direction which can be attributed as the entrance effect in ref. [18]. These phenomena show that both main and secondary channel flows are distributed uniformly. Thus, the flow regimes in the middle region are not influenced by the edge effects. Both main and secondary channel mass flow rates in the filling region ($24<n<30$) are much lower than the middle region except for a large amount of flow mixing at the outlet region due to the flow outlet section area expansion. The fluid flow distribution shows the flow mixing in the filling region is poor except for the filling corner 2. Therefore, it can be seen that edge effects exist along the spanwise location and it may cause hotspots and compromise heat transfer.

Interestingly, the flow distribution for un-blockaded heat sink in Fig. 3.25a is rather uniform along the spanwise location compared to blockaded heat sink in Fig. 3.24a. This phenomenon shows a uniform flow development throughout the full domain un-blockaded cylindrical minichannel heat sink. In the regular cylindrical oblique fin structure, flow mixing is improved between main channel flow and secondary channel flow due to the periodic oblique fin structure. Figure 3.25b, c show the fluid flow distribution in the main and secondary channel for un-blockaded cylindrical heat sink. It is found that the magnitude of main channel mass flow rate is almost constant (0.000346 kg/s) at various spanwise and streamwise locations.

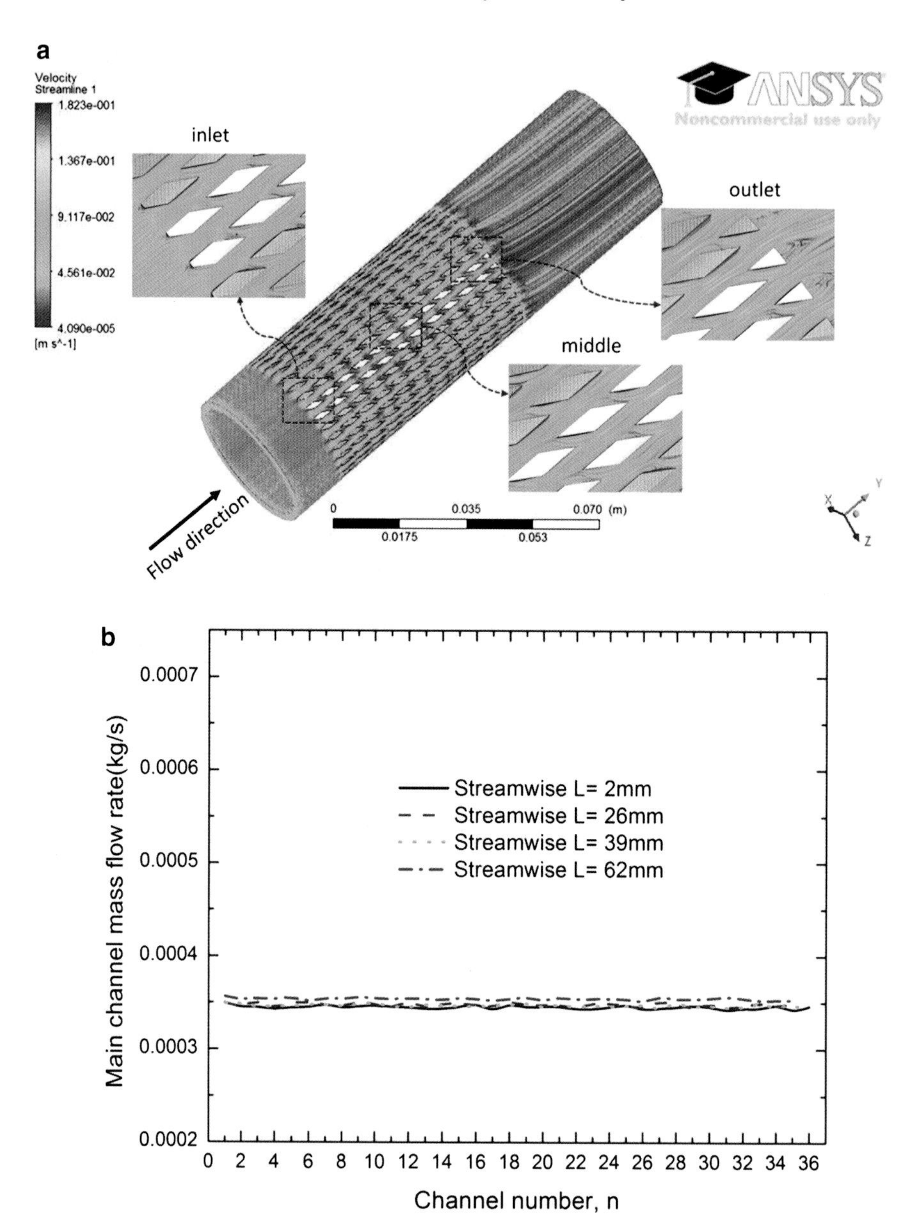

Fig. 3.25 Flow distribution of the un-blockaded cylindrical heat sink: (**a**) velocity contour for flow field domain, (**b**) main channel mass flow rate, (**c**) secondary channel mass flow rate [12]

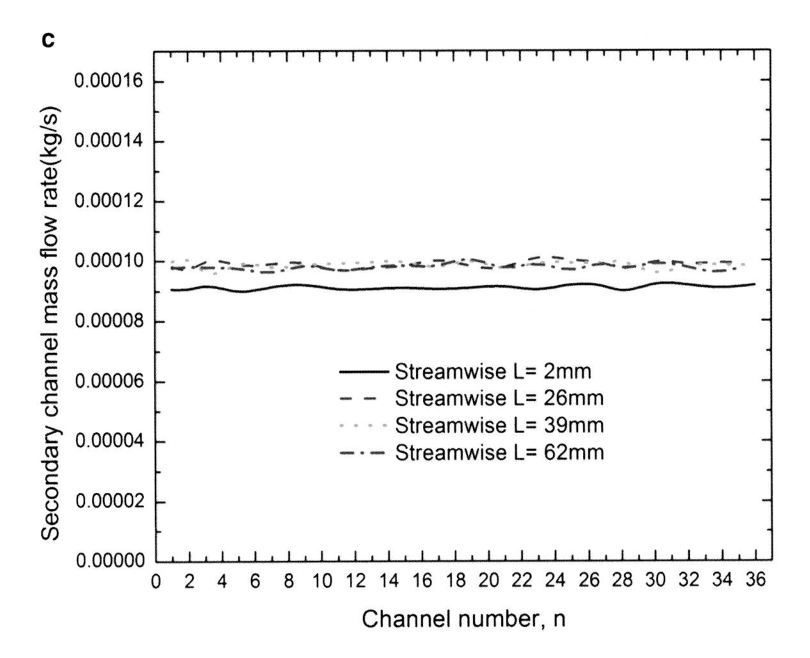

Fig. 3.25 (continued)

There is slight increase at the exit and it is because the main channel mass flow rate at exit is the combination of the secondary and main channel flow. In Fig. 3.25c, the magnitude of secondary flow in the entrance region is slightly lower than the rest of the flow region. One possible explanation for this is that the majority of the flow goes through the main channel and transverse momentum through the secondary channel is very limited in the entrance region. The averaged mass flow rate in the secondary channel is 0.098 g/s, which is around 22 % of the total channel flow. Therefore, it can be concluded that the flow fields for both main and secondary channels are distributed uniformly in the spanwise location in fluid domain. The flow mixing is enhanced and this could result in an improved heat transfer performance.

3.3.2.3 Fluid Temperature Profile

Due to the significant flow field difference, large fluid temperature distinction is found between the blockaded and un-blockaded heat sink in the fluid domain temperature contour in Fig. 3.26a, b. It can be observed that the fluid temperature is relatively uniform at the inlet region. One possible explanation is that the convective heat transfer is more significant at the entrance region and such phenomenon leads to a uniform and lower temperature at inlet region. However, the local fluid temperature profile is non-uniform at the outlet region which is due to flow field distribution as explained previously. Figure 3.26c shows the temperature contour and distribution (mass-weighted average) for channel fluid at the exit location of the

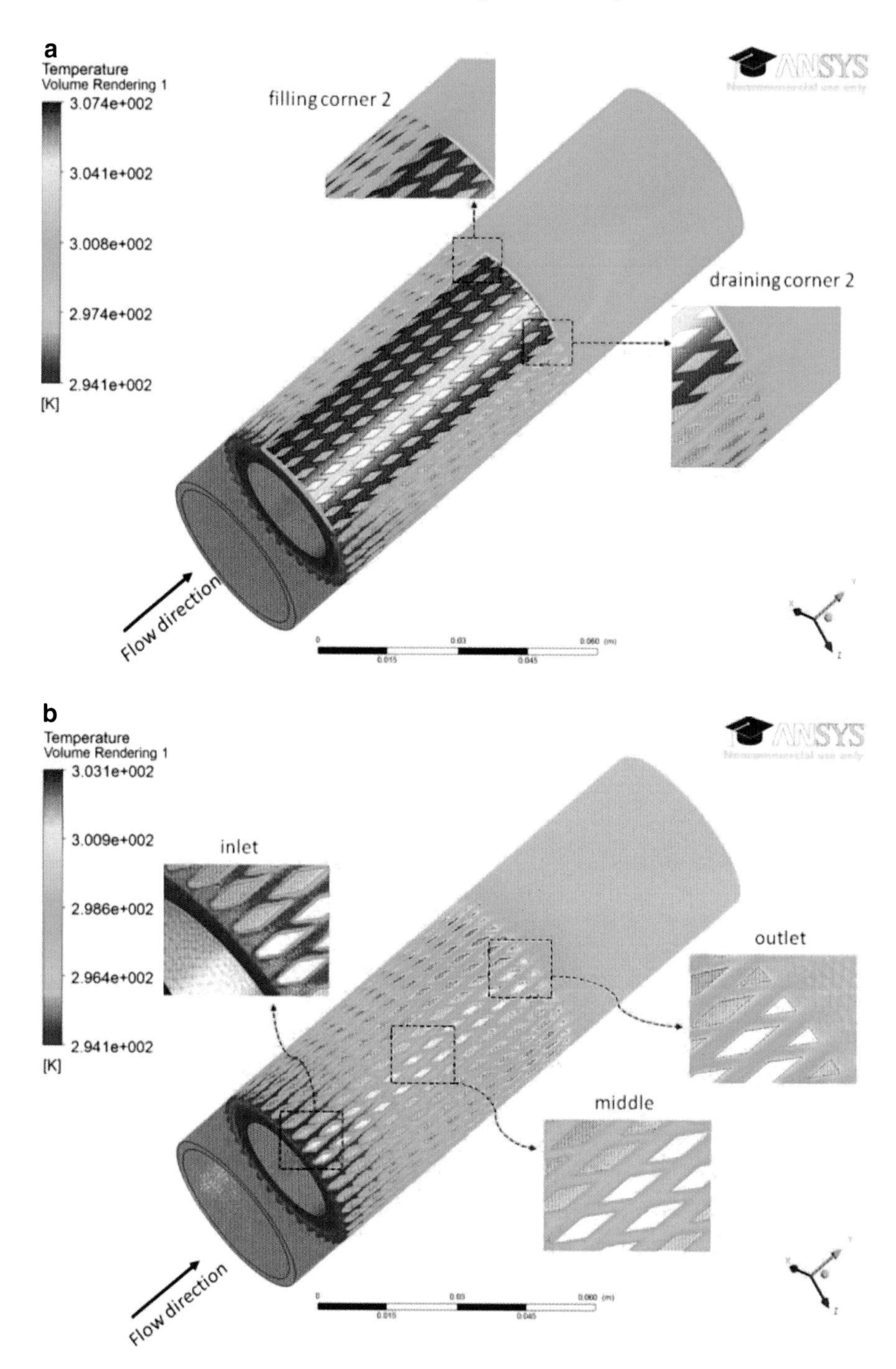

Fig. 3.26 Temperature contour of fluid domain for minichannel heat sink: (**a**) blockaded, (**b**) un-blockaded, (**c**) local fluid temperature comparison at outlet region [12]

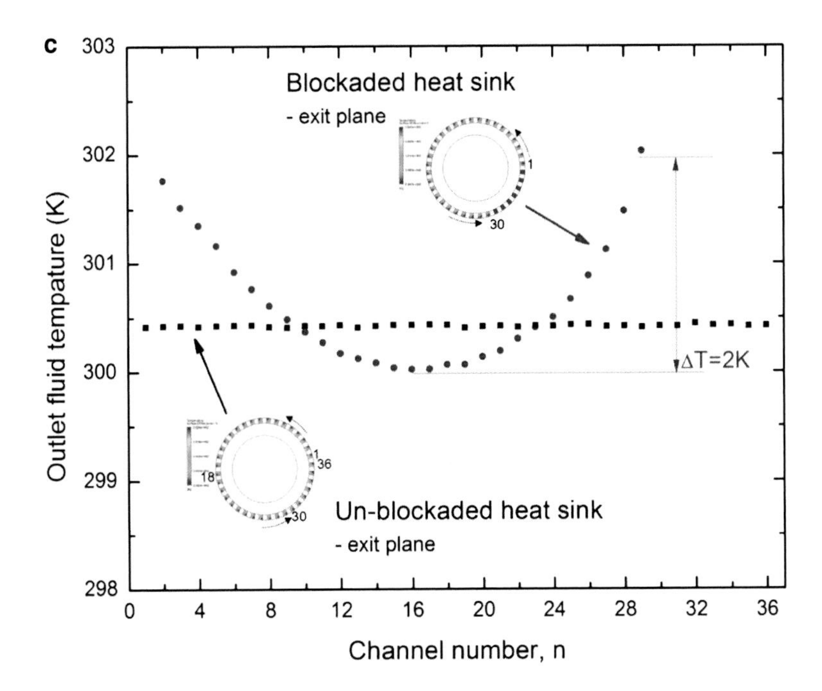

Fig. 3.26 (continued)

heat sinks in the radial direction. It is found that the local fluid temperature for blockaded minichannel at exit region exhibits a concave shape and that for un-blockaded heat sink shows a uniform profile. The maximum temperature gradient for blockaded heat sink is 2 K due to the edge effect. Furthermore, it is observed that a portion of the main flow is diverted into the secondary channel due to the presence of the oblique cuts on the solid fins. This secondary flow, which carries the momentum driven by pressure difference, injects into the adjacent main channel that disrupts the boundary layer and accelerates the heat transfer into the core fluid. This results in the fluid mixing and superior heat transfer performance which leads to lower surface temperature.

3.3.3 Wall Effect and Temperature Uniformity Characteristics

The local spanwise temperature distribution at streamwise locations 26 and 39 mm for blockaded and un-blockaded test piece are presented in Fig. 3.27. The solid line in Fig. 3.27 represents the numerical result of local wall temperature at heat input P of 300 W and mass flow rate Q of 800 ml/min, while the dot symbols plotted in these figures are the experimental data. It is observed that the general trend of numerical results matches well with the experimental results. In Fig. 3.27a, at

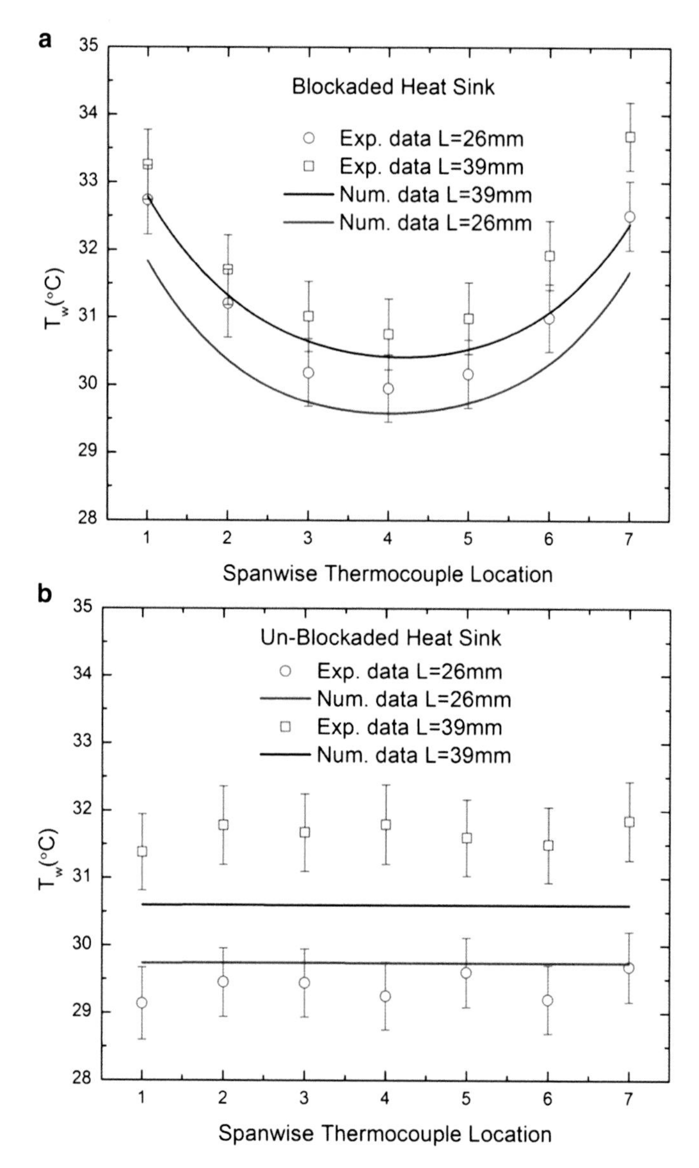

Fig. 3.27 Local spanwise temperature distribution for (**a**) blockaded heat sink, (**b**) un-blockaded heat sink at different streamwise location when heat input is 300 W and mass flow rate is 800 ml/min (experimental and simulation) [12]

streamwise location 26 mm, the deviation between experimental data and numerical data is from 1.1 to 2.7 %, while it is from 0.9 to 3.8 % for streamwise location 39 mm. It is observed that temperature in the spanwise locations 1 and 2 is higher than the middle region. The reason behind this phenomenon is the poor flow mixing caused by lower secondary mass flow rate, which is explained in Sect. 3.4.

For locations 6 and 7, local temperature is higher than the middle region which is because of the combined effect of lower regional secondary mass flow rate and slightly decreased main channel mass flow rate. Interestingly, lower magnitude of local temperature in middle region shows uniform flow mixing between main channel and secondary channel and is not influenced by edge effects. For un-blockaded heat sink, it is observed that the general trend of numerical results agrees well with the experimental results in Fig. 3.27b. The temperatures for both streamwise locations are uniformly distributed. At streamwise location 26 mm, the deviation between experimental data and numerical data is from 0.2 to 2 %, while it is from 2.4 to 3.7 % for streamwise location 39 mm. Thus, it is proven conclusively that cylindrical oblique fin heat sink can achieve a uniform flow distribution and flow mixing due to an absence of edge effect.

Figures 3.28 and 3.29 display the local spanwise temperature distribution at two streamwise locations (26 and 39 mm) when the total heat input increases from 100 to 300 W. Due to the existence of edge effects on the blockaded heat sink, flow field is distributed non-uniformly and this results in hotspot regions on the surface of the heat sink. To express this trend clearly, a quadratic polynomial curve was employed to fit the experimental data. The solid trend line represents the local temperature distribution of un-blockaded heat sink, while the dashed trend line plotted in these figures is the experimental data for blockaded heat sink. The local temperature distribution trend lines for the blockaded test section are shown to be concave shaped and the lowest temperature region is at the middle region of test section. This can be explained by the flow field study in Sect. 3.4. Nevertheless, the temperature fitting curves for un-blockaded oblique fin heat sink are almost flat, and the deviation is reduced when the flow rate increases to 800 ml/min.

These experimental results indicate different temperature distribution uniformity at various mass flow rates for blockaded and un-blockaded heat sinks. To quantify the temperature distribution uniformity, T_{max}–T_{min} is calculated and compared based on the experimental measurements. T_{max} and T_{min} are the maximum and minimum temperatures on the surface of the test locations, respectively, that depend on the location of thermocouples. A smaller value of T_{max}–T_{min} means better uniformity of surface temperature distribution. According to the graphs of T_{max}–T_{min} in Figs. 3.28 and 3.29, the average temperature uniformity of un-blockaded heat sink is always better than that of blockaded heat sink. For the blockaded heat sink, the highest value of T_{max}–T_{min} could reach to 2.94 °C due to the edge effect, while it is only 1.44 °C for the un-blockaded heat sink. In addition, almost all the values of T_{max}–T_{min} for the un-blockaded heat sink are within the error bar at various flow rates compared to blockaded heat sink. This indicates that the surface temperature distribution for un-blockaded heat sink is more uniform than that for the blockaded heat sink. One possible explanation for this is that the flow mixing between the main channel and secondary channel improves and enhances the convection heat transfer performance. Therefore, the cylindrical oblique fin heat sink is an effective cooling method that maintains a uniform surface temperature distribution due to the absence of edge effect.

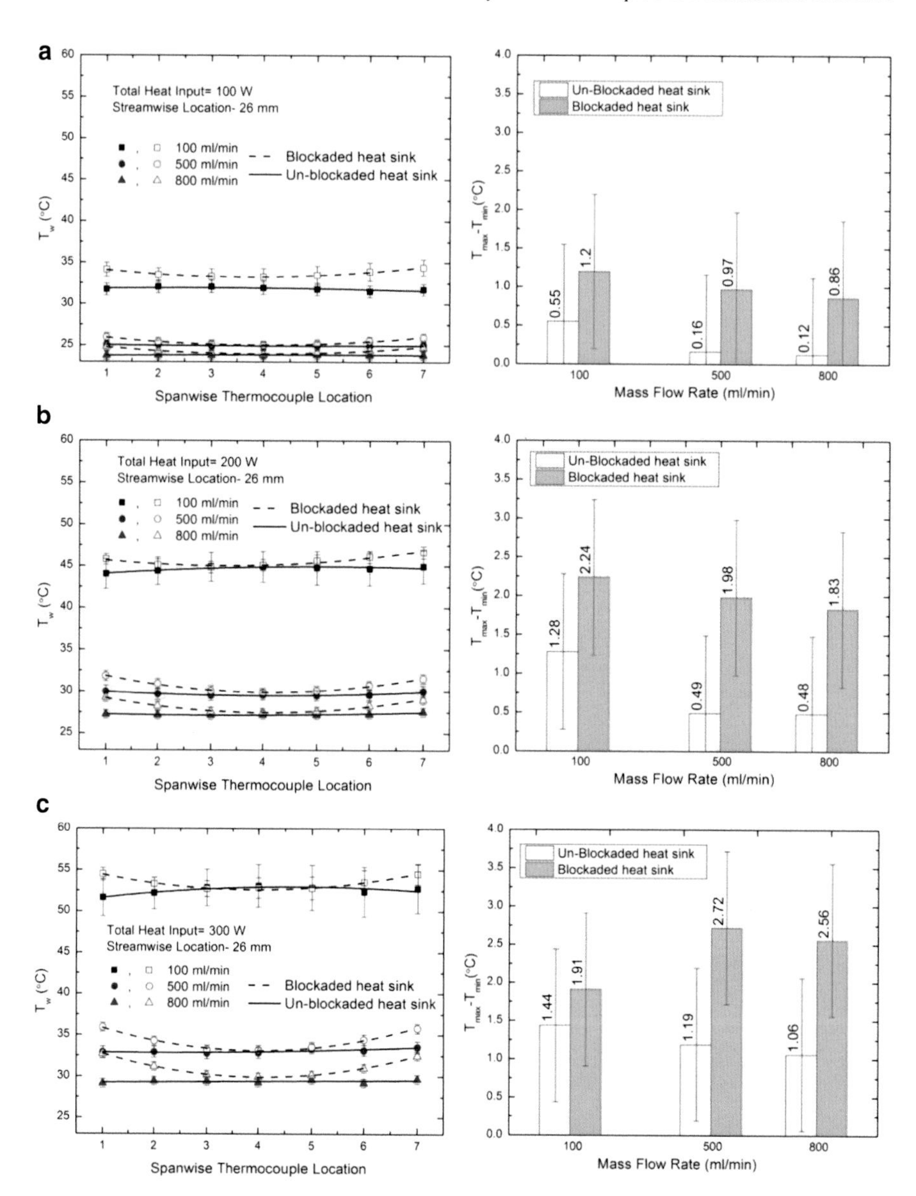

Fig. 3.28 Local spanwise temperature distribution at streamwise location 26 mm from inlet location (**a**) $P = 100$ W, (**b**) $P = 200$ W, (**c**) $P = 300$ W [12]

3.4 Parametric Investigation of Cylindrical Oblique Fin Minichannel Heat Sink

Although the single-phase fluid flow and heat transfer characteristics in cylindrical minichannel with oblique fins are now better understood from previous sections, the fundamental understanding of flow mechanisms in the oblique fin structure is still

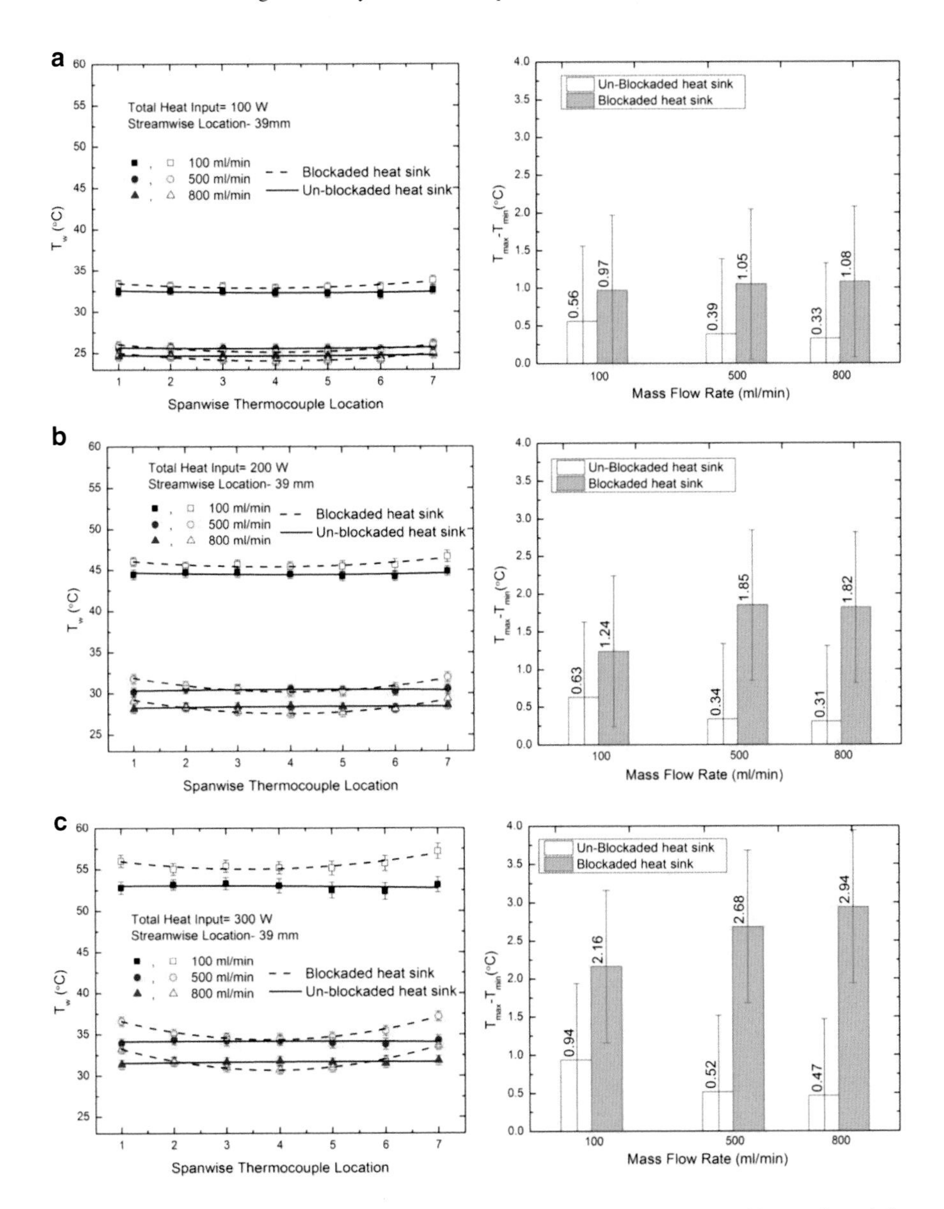

Fig. 3.29 Local spanwise temperature distribution at streamwise location 39 mm from inlet location (**a**) $P=100$ W, (**b**) $P=200$ W, (**c**) $P=300$ W [12]

not fully comprehended. The flow mechanism and optimization of the structure of cylindrical oblique fin heat sink for its best overall heat transfer performance are attempted. A similarity analysis of oblique fin is performed to obtain the dimensionless grouping parameters which are used to evaluate the total heat transfer rate of the heat sink. 3D conjugate heat transfer simulations are carried out using CFD approach to determine the performance of the heat sink. Various flow distributions are investigated and reported, as the secondary channel gap, oblique angle and Reynolds number are varied. Parametric studies were performed by varying the oblique angle from 20° to 45°, secondary channel gap from 1 to 5 mm and Reynolds number from 200 to 900. Key findings are summarized as follows [19]:

1. Similarity analysis of oblique fin structure was firstly performed by employing wedge flow scheme. Boundary layer analysis showed that the average Nusselt number is a function of five independent variables, β, l_{sc}/l_u, D_h/L, Reynolds number and Prandtl number.
2. The flow distribution as l_{sc} increases was investigated. As l_{sc} increases from 1 to 3 mm, the flow is main channel directed, and the recirculation zone in secondary channel increases along the flow direction. As l_{sc} increases from 3 to 5 mm, the flow pattern becomes more and more secondary channel directed and recirculation moves upwards and disappears at the downstream by compromised heat transfer area reduction.
3. The flow distribution as oblique angle increases was also examined. The amount of secondary flow increases and the boundary layer is disrupted as oblique angle is increased from 20° to 30°. However, the recirculation zone also increases correspondingly and deteriorates the convective heat transfer as oblique angle is increased from 30° to 45°.
4. The flow distribution as Reynolds number increases was analysed for oblique fin minichannel ($\theta = 30°$, $l_{sc} = 2$ mm). The boundary layer keeps redeveloped at each leading edge of oblique fin, and this enhances the heat transfer performance at a Reynolds number of 526. As Reynolds number increases to 900, the recirculation zone is intensified, which hinders the heat transfer and suffers a high pressure drop penalty.
5. The increasing trend of average Nusselt number decreases when Reynolds number is more than 500 at large oblique angle ($35° < h < 45°$) structure due to recirculation generated at larger oblique angle, and the increasing trend of average Nusselt number decreases greatly at lsc of 5 mm at large oblique angle structure due to the reduced heat transfer area which deteriorates convective heat transfer and increases pressure drop penalty.
6. Based on 259 numerical data points, multiple correlations for the average Nusselt number and apparent friction constants in terms of appropriate dimensionless parameters were obtained. These correlations successfully pave the way for optimization of the oblique fin heat sink without the need of numerical simulation analysis or fabrication of the heat sink.

References

1. Fan Y, Lee PS, Jin L-W, Chua BW (2013) A simulation and experimental study of fluid flow and heat transfer on cylindrical oblique-finned heat sink. Int J Heat Mass Transfer 61:62–72. doi:10.1016/j.ijheatmasstransfer.2013.01.075
2. Lee YJ, Lee PS, Chou SK (2010) Enhanced microchannel heat sinks using oblique fins, ASME 2009 InterPACK Conference, 19–23 July, vol 2. San Francisco, CA, USA, pp 253–260. doi:10.1115/InterPACK2009-89059.

3. Suga K, Aoki H (1995) Numerical study on heat transfer and pressure drop in multilouvered fins. J Enhanc Heat Transfer 2:231–238
4. DeJong NC, Jacobi AM (2003) Flow, heat transfer, and pressure drop in the near-wall region of louvered-fin arrays. Exp Therm Fluid Sci 27:237–250
5. Satish G, Kandlikar SG, Garimella S, Li D, Colin S, King MR (2005) Heat transfer and fluid flow in minichannels and microchannels, 1st edn. Elsevier, Oxford, UK
6. Shah RK (1978) Laminar flow forced convection in ducts: a source book for compact heat exchanger analytical data. Academic, New York, NY
7. Alharbi AY, Pence DV, Cullion RN (2003) Fluid flow through microscale fractal-like branching channel networks. J Fluid Eng Trans ASME 125:1051–1057
8. Fan Y, Lee PS, Jin L-W, Chua BW (2014) Experimental investigation on heat transfer and pressure drop of a novel cylindrical oblique fin heat sink. Int J Therm Sci 76:1–10. doi:10.1016/j.ijthermalsci.2013.08.007, http://dx.doi.org
9. Kandlikar SG, Li D, Colin S, King MR (2005) Heat transfer and fluid flow in minichannels and microchannels. Elsevier, Oxford
10. Rosaguti NR, Fletcher DF, Haynes BS (2006) Laminar flow and heat transfer in a periodic serpentine channel with semi-circular cross-section. Int J Heat Mass Transfer 49:2912–2923
11. Bennett D (2013) Why are the batteries in Boeing's 787 burning? Internet article-http://www.businessweek.com/articles/2013-01-18/why-the-batteries-in-boeings-787-are-burning
12. Fan Y, Lee PS, Chua BW (2014) Investigation on the influence of edge effect on flow and temperature uniformities in cylindrical oblique-finned minichannel array. Int J Heat Mass Transfer 70:651–663. doi:10.1016/j.ijheatmasstransfer.2013.11.072, http://dx.doi.org
13. Rhie CM, Chow WL (1983) Numerical study of the turbulent flow past an airfoil with trailing edge separation. AIAA J 21:1525–1532. doi:10.2514/3.8284
14. Sparrow EM, Abraham JP, Minkowycz WJ (2009) Flow separation in a diverging conical duct: effect of Reynolds number and divergence angle. Int J Heat Mass Transfer 52:3079–3083
15. Menter FR (1994) Two-equation eddy-viscosity turbulence models for engineering applications. AIAA J 32:1598–1605
16. Lee GG, Allan WDE, Goni Boulama K (2013) Flow and performance characteristics of an Allison 250 gas turbine S-shaped diffuser: effects of geometry variations. Int J Heat Fluid Flow 42:151–163
17. Bardina JE (1997) Turbulence modeling validation, testing, and development [microform]/ J.E. Bardina, P.G. Huang, T.J. Coakley National Aeronautics and Space Administration, Ames Research Center; National Technical Information Service, distributor, Moffett Field, CA; Springfield, VA
18. Lee YJ, Lee PS, Chou SK (2013) Numerical study of fluid flow and heat transfer in the enhanced microchannel with oblique fins. J Heat Transfer 135(4):041901
19. Fan Y, Lee PS, Jin L-W, Chua BW, Zhang D-C (2014) A parametric investigation of heat transfer and friction characteristics in cylindrical oblique fin minichannel heat sink. Int J Heat Mass Transfer 68:567–584. doi:10.1016/j.ijheatmasstransfer.2013.09.027, http://dx.doi.org

Chapter 4
Conclusions and Future Works

4.1 Conclusions

A novel heat transfer enhancement scheme is proposed, employing oblique fins to replace the conventional parallel fin in a micro-/minichannel heat sink. The following key conclusions are drawn from this study:

- Oblique fin design modulates flow, creating a uniquely skewed and asymmetrical velocity and temperature profiles. Such profiles result in thinner boundary layer thickness. The presence of oblique channel also diverts a fraction of coolant into it and generates secondary flow to further enhance the fluid mixing. The combination effect of thinner boundary layers and secondary flow generation causes the flow to maintain a developing state leading to a great improvement and uniform heat transfer performance. It is also noticed that the level of heat transfer enhancement is closely related to the secondary flow rates through oblique channel. The combined effect of redevelopment of boundary layer and generation of secondary flow results in thinner boundary layer that leads to better heat transfer performance and higher pressure drop. The increased pressure drop was somehow compensated by pressure recovery due to the "diffuser" effect at each diverging section.
- Heat transfer performance of copper-based enhanced microchannel heat sink with 500 μm nominal channel width is greatly augmented as compared with the conventional microchannel. For the lowest total flow rate which corresponds to Re ~300, the minimum heat transfer enhancement factor is close to 1.6. When the Reynolds number is increased beyond 600, heat transfer enhancement could reach up to 2.2.
- For the case of enhanced microchannel with 500 μm nominal channel width, there is negligible pressure drop penalty for the heat transfer enhancement achieved up to Re ~400, which is the most attractive feature of this scheme. Although further increment in Reynolds number will incur additional pressure

Y. Fan et al., *Thermal Transport in Oblique Finned Micro/Minichannels*,
SpringerBriefs in Applied Sciences and Technology,
DOI 10.1007/978-3-319-09647-6_4

drop penalty, the heat transfer enhancement factor is further increased and it is always larger than the pressure drop penalty. As for enhanced microchannel heat sink with 300 μm nominal channel width, burr/bend at the fin edge is suspected to result in higher pressure drop. Unless better fabrication technique is used in fabrication, there is always risk of having some degree of burr for enhanced microchannel heat sink with fin width less than 500 μm.

- Experimental demonstration through two hotspot scenarios shows that the positioning of finer fin clusters is flexible and regardless of hotspot location on the chip. For both hotspot scenarios experimentally tested, variable pitch oblique-finned heat sink achieves the lowest maximum temperature and temperature gradient among three microchannel heat sink configurations. The reduction in maximum temperature and temperature gradient can be as high as 17.1 and 15.3 °C, respectively. As the remaining chip area is not overcooled, the resultant chip temperature contour is much more uniform. Also the heat transfer enhancement can be achieved with affordable pressure drop penalty, estimated at 10–20 %.
- Better fluid mixing and superior heat transfer performances are achieved by the secondary flow generation for cylindrical oblique fin minichannel. A flow recirculation zone forms at larger Reynolds number in the secondary channel. However, this recirculation is insignificant in the low Reynolds number study. Both hydrodynamic entrance length L_h and thermal entrance length L_t in cylindrical oblique fin minichannel were found much shorter compared to conventional straight fin.
- The local Nusselt number for the cylindrical minichannel with oblique fins nearly reaches a constant value of 11.8 along the downstream direction, which is almost 90 % enhancement compared with conventional minichannel. Surfaces *O1* and *M1* were found as the most effective heat transfer surface, while *BO2* is the most ineffective heat transfer surface. For a heat flux of 6.1 W/cm^2 and Reynolds number of 310, local wall temperature distribution showed that the surface temperature of cylindrical heat sink is reduced by 4.3 °C overall. Heat transfer enhancement (E_{Nu}) and pressure drop penalty (E_f) showed significant advantages of the cylindrical oblique fin minichannel over the conventional straight fin minichannel. Based on 259 numerical data points, multiple correlations for the average Nusselt number and apparent friction constants in terms of appropriate dimensionless parameters were obtained. These correlations successfully pave the way for optimization of the oblique fin heat sink without the need for numerical simulation analysis or fabrication of the heat sink.
- Numerical simulation studies for full domain cylindrical oblique fin heat sink were validated by the experimental tests. Edge effect can cause the formation of localized hotspots due to the non-uniform flow distribution in the planar oblique fin configuration. On the contrary, the cylindrical oblique fin can periodically induce the hydrodynamic boundary layer development to reinitialize in both spanwise and streamwise directions. This could result in better fluid mixing and superior heat transfer performance. The flow field study showed that poor flow

mixing exists in the draining and filling regions, while the flow regime in the middle region is not influenced by the edge effects for the blockaded cylindrical oblique fin heat sink. The flow fields for both main and secondary channels are distributed uniformly in the spanwise location for regular cylindrical oblique fin heat sink. The local temperature distribution trend lines for blockaded cylindrical oblique fin heat sink are shown to be uniquely concave shaped due to the edge effect. However, a more uniform and lower surface temperature distribution for regular cylindrical oblique fin heat sink is observed as a result of the improved flow mixing without any edge effect.

4.2 Future Works

Encouraged by the achievements of present numerical and experimental investigations, the major extended works for future research are recommended and outlined as follows:

- Investigate the effect of staggered oblique fin design. The present study adopts conventional manufacturing techniques in fabricating the test vehicles and limits the oblique fin designs to only in-line arrangement for main channel and oblique channel. On the other hand, initial trials with numerical simulation show staggering cut in oblique channel could result in greater secondary flow generation and thus higher heat transfer performance. For this, a more flexible fabrication technique such as micro-fabrication can be employed.
- Include a turnaround oblique fin in the design consideration, which is similar to turnaround louvre in louvred fin heat exchanger. The turnaround oblique fin aims to flip the direction of secondary flow for the bottom half of the heat sink and reduces the net flow migration across the heat sink to zero.
- Critical Reynolds numbers could be identified for oblique fin structure. The numerical analysis with appropriate viscous models could be used and validated by numerous experimental measurements. At a higher Reynolds number region, various flow mechanisms such as chaotic eddies, flow instabilities and channel surface roughness may be detected.
- In order to get an improved physical insight into fluid flow and friction characteristics for the present heat sink, flow visualization studies could be performed using dye injection method or micro PIV (particle image velocimetry). With the results, recirculation zone, flow mixing enhancement, secondary flow percentage and the edge effect could be quantified and validated with the numerical studies and experimental measurements.
- The present work focuses on single-phase liquid cooling and the coolant is only water. In order to develop a more complete understanding of oblique-finned heat sink, heat transfer and pressure drop data for different fluids and substrate materials should be generated. Air cooling, flow boiling and two-phase flow in oblique fin minichannel could also be investigated further.

- Based on the proposed correlations for cylindrical oblique fin heat sink, a more generalized correlation could be derived based on the height of the channel (H), the total length of the channel (L), oblique fin angle (θ), aspect ratio (α), oblique fin length (a), secondary channel gap length (b) and Reynolds number (Re). Using this algorithm, various optimized oblique fin designs can be obtained in different applications. Experiments can be carried out based on the optimized geometries under different mass flow rates to verify the algorithm.